The Complete
Jewelry Making Course

The Complete
Jewelry Making Course

Jinks McGrath

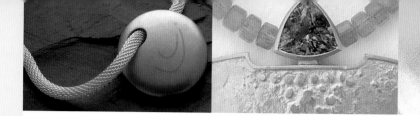

A QUARTO BOOK

Copyright © 2007 Quarto Publishing plc

First edition for North America
published in 2007 by
Barron's Educational Series, Inc.

All inquiries should be addressed to:
Barron's Educational Series, Inc.
250 Wireless Boulevard
Hauppauge, NY 11788
http://www.barronseduc.com

Library of Congress Control Number: 2006936734

ISBN-10: 0-7641-3660-7
ISBN-13: 978-0-7641-3660-3

Conceived, designed, and produced by
Quarto Publishing plc
The Old Brewery
6 Blundell Street
London N7 9BH

QUAR. CJM

Senior Editor: Liz Dalby
Copy Editor: Natasha Reed

Art Director: Caroline Guest
Art Editor: Julie Joubinaux
Designers: Balley Design Ltd

Photographers: Paul Forrester and Phil Wilkins
Picture Researcher: Claudia Tate

Creative Director: Moira Clinch
Publisher: Paul Carslake

Manufactured by Modern Age Repro House Ltd, Hong Kong
Printed by Toppan Leefung Printing Ltd, China

19 18 17 16 15 14 13 12 11 10

Contents

Chapter 3: Practice projects — 124

Introduction

During the past 35 years of making, teaching, and writing about jewelry making, I don't think I have spent even one day in the workshop when I haven't learned something new, tried a different way of doing something, or talked with colleagues about their way of making something. This is, I think, the reason why every day continues to be such a pleasure and the reason I continue to have a consuming interest in all aspects of jewelry making.

In this book I go back to the beginning, not only to explain how to do something but to explain why it might work one way and not another. I also give advice on the correct use of tools and the importance of using the right tool for the job. I answer the questions that I found myself asking and puzzling about all those years ago!

Jinks McGrath

About this book

The *Complete Jewelry Making Course* is arranged into 34 units, covering the essential aspects of jewelry making. Six practice projects draw together the skills you have learned.

Chapter 1: Getting started

Find out where to look for inspiration, how to develop ideas, and about the tools and materials you need to start making jewelry.

TECHNICAL INFORMATION
Technical information is clearly presented.

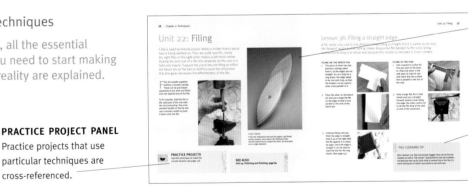

ILLUSTRATED PANELS
Illustrated panels highlight key information or ideas.

"SEE ALSO" BOX
"See also" boxes suggest links to related units.

Chapter 2: Techniques

In this section, all the essential techniques you need to start making your designs reality are explained.

PRACTICE PROJECT PANEL
Practice projects that use particular techniques are cross-referenced.

INSPIRATIONAL EXAMPLES
Throughout the book, specific teaching points are illustrated by photographs of inspirational examples.

TIPS
Helpful hints and tips are provided to help you get the best out of your tools.

Chapter 3: Practice projects

These six projects are specially designed to test the skills covered in the techniques chapter.

MATERIALS AND TOOLS
The materials and tools needed to complete the project are listed in the order they are used.

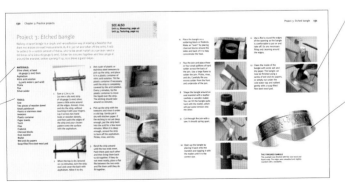

STEP BY STEP INSTRUCTIONS
Each lesson or project is explained by clear text and informative photographs.

PHOTOGRAPH OF THE FINISHED PIECE
A clear photograph of the finished piece shows what you should be aiming to achieve.

chapter 1
getting started

This chapter looks at different ways of finding inspiration and understanding how to use your ideas in the most practical way to suit your skills. Take every opportunity available to look around you, sketch ideas, and be stimulated by anything and everything. Ideas may start to flow only after you have had some practical experience with metal and tools. If so, take some time out between your practical sessions or classes to find out what inspires you, so that when you next sit down at the bench you have an idea of where you are heading.

Unit 1: Finding inspiration

Finding out what pleases you and what doesn't is the first task in this unit. With anything related to design, it is what you see—what inspires and excites you—that is important. The ideas on these pages are just a guide to help you find the methods that work best when you start to create your own ideas and designs.

Visits to museums, galleries, and exhibitions are an excellent way to begin. Even if you live far away from a major city with national museums, there is plenty of inspiration to be found by visiting your nearest town's own museum. Fragments of old pottery or glass displayed there can be tremendously interesting, as are images of local industry—whether it has been a major influence on the area in the past, or is still.

If you are fortunate enough to have access to a larger museum, then seek out the specialist jewelry section. You could also look at the metalwork and blacksmithing sections, which tend to house artifacts relevant to jewelry making. Consider also armor, swords, knives, tea and coffee sets, flatware, and stained glass such as Tiffany lamps.

PLAN AND PREPARE

Try to assess what you want to see before setting off on a long trip around a museum. Ask for a map of the layout and visit the rooms you are interested in first. Take a notebook and pencil with you and jot down anything that you find truly beautiful.

LOOK CLOSELY

When you examine a piece of jewelry, have a really close look and try to decide how it might have been made. Look for repairs and seams, or ask the museum curator if an item has been X-rayed. This could help you figure out how it was put together.

ADAPT IDEAS

Be inspired by the photographs of jewelry in books, but try to bring something of your own to any design idea that you have as a result. For example, try simplifying an idea by retaining only the shape or color of the original piece.

Gathering ideas

Get into the habit of gathering and recording inspirational ideas.

POSTCARDS

Start a postcard collection of pieces that you have seen and liked from exhibitions. Pin them on a wall or bulletin board to provide something inspirational to look at with your first cup of coffee in the morning.

SCRAPBOOK

Cut out pictures from magazines of things that interest you and keep them in a scrapbook or album. Or scan and download images and use your computer to help you file them all.

WHERE TO LOOK

Museums: Visit museums and study the jewelry, brass and copperware, agricultural and industrial tools, and anything else that you find stimulating. Remember that small local museums can be fascinating and will give you an excellent feel for the topography of an area.

Galleries: Find out where your local galleries are. Visit all their exhibitions and ask to be put on their mailing list—these resources are there for people like you. Recognize which exhibitions you enjoy and those which you don't. It's fine not to like them all!

Exhibitions and open studios: Read local newspapers or listings in magazines for news of other exhibitions. For example, your town may have a festival where local artists open their studios to the public. Take the opportunity to see other artists' work, buy inspirational pieces, and talk to them about what they do. Some exhibits and fairs may feature artists actually making their work. Take the opportunity to observe the techniques and tools they use.

Magazines: Most magazines connected with fashion carry jewelry advertisements or even special features about jewelry. Even in magazines that are unrelated to fashion, you may find pictures of people wearing jewelry or lists of galleries exhibiting jewelry. There are specialist magazines too—ask at your local library or news store to find out what's available.

Books: Books about jewelry provide a fantastic insight into the way jewelers around the world work. Research these sources of inspiration in your local library, or look up jewelry-related books online.

The Internet: Use the image option in a search engine to find inspirational and informative photographs, drawings, and graphics related to any subject you can think of from a huge range of sources. As with any source material, don't copy other artists' work, but do use it to inspire you.

INSPIRATION EXERCISE

Try the following exercise to discover a method of finding inspiration that is unique to you. There are no hard-and-fast rules. It is a journey of discovery to find the things that have some kind of emotional effect on you: things that appeal to you no matter what the reason; things that you find interesting, beautiful, controversial, dark, or exciting; things that can open up ways of seeing everyday objects in a totally different light.

Wherever you live, take time out for a walk. Look all around you—up and down. What shapes can you see outlined against the sky? What is level with your shoulder? What sort of ground are you walking on? What is the road layout? Maybe it is a path through the woods; maybe a trip to the mall—wherever you are there will be something of interest to you if you look closely. How does the light throw a shadow? Can you see reflections in glass? How does a raindrop hang on a leaf? Is there interesting ironwork as you walk through the park? Even if you can't use any of the things you observe directly, just being aware of their influence will have an effect on your ideas.

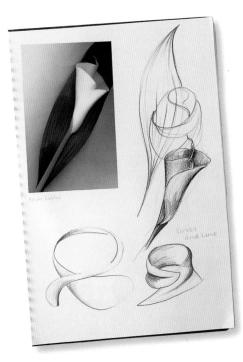

CURVING FORMS

Here, the jeweler used a photograph of a lily for inspiration. The long curves of the flower are translated into a beautiful twist, giving ideas for a ring or maybe a pendant.

WHAT TO LOOK AT

Light and dark: Change the way you look at shadows and the way light reflects on surfaces; try to reduce them into distinct shapes. Watch how the shape of a particular shadow moves during the day and becomes elongated. Really look at horizons and silhouettes and see the different shades and shapes of dark there are.

Buildings: Architecture can be a wonderful source of ideas. Some modern buildings have the most fantastic physical forms and outlines. Many older buildings, such as cathedrals, possess a unique sense of atmosphere, a legacy of the almost unbelievable skill and dedication of the architects and builders who created them.

Nature: What great painter, sculptor, potter, jeweler, or designer has not at some point been inspired by the natural world? The way the center of each flower is so perfect, the way the sun sets each evening, and the way the moon intrigues us night after night. Let yourself be influenced by amazing things in which humans have had no part in making.

Fashion: As a designer and jewelry maker it is highly appropriate to be aware of fashion. Be excited by trends in new colors and shapes on the catwalk. You don't need to be led by what is fashionable at any one time, but there is a great deal to be learned by observing what is going on "out there."

World cultures: Open yourself to influences from a range of cultures. For example, traditional African textiles, indigenous Australian art, or Pacific island costumes and masks are all fabulous sources of ideas that can inform your designs.

Exploring ideas

Learn how to analyze what you see in your search for inspiration. Transpose ideas rather than taking them literally. Play with materials and effects to create something new and original.

LIGHT AND DARK

Create a design from the ideas around light and dark. Cut out shapes from thick black paper and place them onto a white sheet to show the contrasting tones and shapes. The shape of the black paper can then be cut again and spread out to create a more interesting form.

FABRIC EFFECTS

Working up an idea from fabric can translate into lovely effects in metal, but remember that metal is not as flexible as fabric. Observe how a full skirt hangs, or is layered, or has an uneven hem. You could also think of using some fabric as an integral part of your design.

SOLID SHAPES

Use solid objects such as buildings to inspire a design. Find photographs of buildings and use pencil and paper to make tracings. Draw the spaces in between the buildings to see what sort of shapes emerge; see the outline they make against the sky.

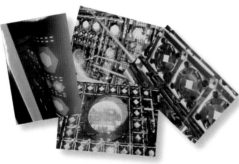

NATURAL OBJECTS

Gather a collection of leaves, stones, flowers, shells, and bark together and keep them together on or around your workbench. Instead of using the whole leaf or flower for inspiration, try dividing it up and working with just a small section; it could be more interesting than trying to copy it entirely.

Unit 2: Using photography

Photography is useful to you as a jewelry designer in two main ways. First, use it to capture anything at all that you find visually inspirational. Keep these images so that you can refer to them whenever you need to. Second, you can photograph your own work, either for your own records or to market it, for example through a website.

When looking for inspiration it's a good idea to keep the camera with you all the time, ready to take a picture of anything that catches your eye—close-ups of a flower or a mosaic tile pattern, for instance. Photography has changed considerably in the last few years and there are now more options than ever. The advantages with a digital camera are that you can get close up to your subject without having to alter the settings, and that you can just keep snapping until you know you have the right shot. From your photos, you may find interesting outlines, light and dark shapes, or perhaps some unusual lettering from old advertisements. Inspiration can come from surprising sources. For example, surface effects such as weatherworn brick or muddy tire tracks may inspire thoughts of beautifully textured silver.

ORGANIC TEXTURES
Close-up photographs of leaves and wood textures were inspirational in the making of this enameled "poppy" necklace with the leaves draped around the solid choker.

DELICATE EFFECTS
A photograph taken through trees has inspired this beautiful necklace. The moonstones seem to capture the delicacy of the sky seen through the dense forest.

SEE ALSO
Unit 4: Translating ideas, page 18
Unit 13: Transferring patterns, page 50

PHOTOGRAPHING YOUR OWN WORK

Choose your lens: A macro lens with automatic focus is essential for photographing jewelry with a handheld camera, because it will allow you to take close-ups. If you use a tripod, the focusing can be done manually.

Plan shots with care: Take care when photographing reflective metal jewelry. Choose a background surface that will complement the piece but not clash with it. If you are taking the photograph from directly overhead, a flat surface should be sufficient. If you plan to shoot the photograph at an angle, you need to curve the surface upward behind the piece.

Take a range of shots: Get in close to your subject and take shots from different angles.

Build a permanent setup: There will be many times when you will want to make your own records of jewelry designs, so consider building a permanent, dedicated setup where you can photograph pieces of jewelry as you finish them. Try a variety of lighting techniques.

Try diffused light: Diffusing the light source minimizes shadows. To achieve this, make a wooden box frame and completely cover it with white tissue paper. Cut a hole for the lens in the top. Put the jewelry inside the box. Position lights at each side of the box and above it. Make sure none of the lights are in direct contact with the paper. Place the camera at the top of the box, with the lens looking down through the hole onto the jewelry.

Try natural light: Natural light will cast shadows, but these can be used to great effect. Place the jewelry on a suitable background, and photograph it outside, either in full sunlight or on a bright cloudy day. Photograph at high noon, in the shadow of a building. This should help to prevent overbright photos or "hot spots."

Try a lightbox: You can buy small lightboxes especially designed for photographing jewelry. Check that the size you want will suit all your needs and ensure that you have seen the typical results before investing in one.

DIGITAL CAMERA SETUP
The digital camera is set up on a tripod. White card is gently folded up behind the piece to give a consistent background, and the jewelry being photographed is set on a gray piece of slate to give a good contrast.

PRACTICE PROJECTS

Use photography to record your finished projects. See pages 126–139.

DESIGNS FROM PHOTOGRAPHS
You may want to transfer the outline of a photographic image onto a sheet of metal for piercing (see page 50). Get a print of the picture you plan to use, in the desired format and at a suitable size. For digital photos, use imaging software to manipulate the image if you need to. Identify the area of the photo that you want to use in your design; choose an area with strong, bold outlines for the best results. Use tracing paper and a pencil to trace around the outlines. The tracing can then be transferred to the metal ready for piercing (see page 51).

Unit 3: Evaluating ideas

Once you have found your inspiration, you need to find a way of translating it into a piece of jewelry. Evaluate each idea you have at this stage: identify which will work, which you can save for future projects, and which have no potential after all.

DEVELOPING IDEAS

Save an idea for later: Try not to get stuck on an idea if it doesn't seem to be going anywhere. You can always go back to it later. If it is a worthwhile idea, you may find your way through it while you are thinking about or working on something completely different.

Backtrack: Don't be afraid to go backward in the design process. For example, you might be cutting up some paper to make a pattern and find that it is not working however much you move the pieces around. Start again! Cut the paper more simply and see where that leads.

Be appropriate: Natural, "found" objects may be so beautiful that it is tempting to think they can be reproduced in metal. Sometimes this simply is not possible. Delicate objects formed in metal can look heavy and undefined—try to imagine realistically how your idea would look.

Draw freely: Make your early drawings as free and as large as you want to. At this stage absolutely nothing needs to be accurate. Your ideas can all be refined as you start to work up an actual design from them. Don't worry about the size or scale of an object when you are first designing. You can always make ideas larger or smaller with a copy machine or by scanning them into the computer.

Using the sources of inspiration you have gathered (see Units 1 and 2, pages 10–15), you should now be starting to come up with starting points for potential projects. At this stage, try making a collection of pieces based around an emerging theme. Find a suitable space, such as a large tabletop, where you can gather materials together to develop your ideas on paper.

Aim for a mixture of natural objects, bits of paper and card, drawing paper, pencils, crayons and paints, printouts, photographs, books, postcards, and sketches. You may also have an opinion on whether you plan to use any beads, stones, or other materials in a finished piece, so have them around as well. Be as adventurous as you like to encourage a free flow of ideas.

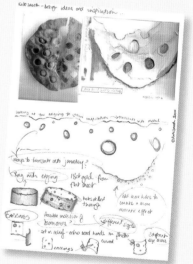

PROGRESSION OF IDEAS
Rough ideas are sketched out, using a seed head as inspiration. The holes in the seed head become the focus of the design, which then becomes a ring.

THE FINISHED RING
The gold ring has been given a slightly rough texture, and the idea of the holes has been used to great effect with diamonds scattered among them.

Making sketches

Begin sketching rough ideas and gradually refine them until you have a clear vision of the finished jewelry. This helps you to plan the piece in a three-dimensional way.

FINISHED PIECE
The finished silver pendant reflects ideas and themes worked out in the development sketches (below).

EXPERIMENT WITH DRAWINGS
Start drawing; experiment with more than one idea. You will begin to see how the jewelry might take shape. Color your work if it is appropriate to the finished piece.

DEVELOPMENT SKETCHES
A succession of drawings on a theme begins to evolve into more finished ideas for a piece of jewelry based on the intricate three-dimensional forms of flower petals and stamens.

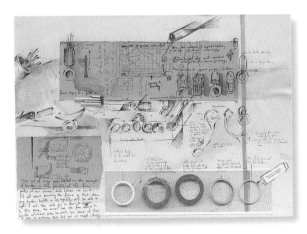

THINK IN THREE DIMENSIONS
Try to think three dimensionally; develop a feeling of shape and form. Make a model from thicker card stock if it helps you.

Unit 4: Translating ideas

All the major decisions about a piece of jewelry are made at the stage when you translate your initial ideas into a working design. There are many things to consider: materials, dimensions, colors, flexibility, wearability, weight, cost, and construction. As you start jewelry making, there may be limitations on how you construct a piece, but as you become more skilled you will be able to consider many more options.

CONSTRUCTION METHODS IDEAS FILE

Smooth and flat: The piece can be cut from a flat sheet with a saw; it will be the same thickness throughout. The saw can also be used to cut any decorative patterns.

Curved: Sheet metal can be annealed (softened) and then curved or shaped by the use of different stakes and metal or wooden blocks.

Three dimensional: If the piece has some areas that will be thicker than others, it can be made using any casting technique or thicker sheet metal and wire.

Patterned surface: A patterned surface can be applied to the sheet metal before it is cut out or shaped, using a rolling mill, hammers, or punches.

Decorative surface: Wires or smaller pieces of metal can be applied to sheet metal, before cutting or after shaping.

Fittings and findings: The fittings and findings are the ear posts or clips, pin fittings, chain fastenings, jump rings, toggles, and so on. These are usually the last pieces to be added to any other type of construction.

SEE ALSO
Unit 12: Measuring, page 46
Unit 20: Joining, page 70
Unit 27: Fittings, page 92

The most important consideration is how a piece will be constructed. This means that you must have an understanding of how you want it to look and feel when it is finished. Ask yourself: How will it hang? How heavy will it be? Where will the fittings be placed to get the correct balance? How will it fasten? All these issues will need to be planned into your design.

ORDER OF CONSTRUCTION
As well as the methods of construction, you need to think about the order in which they will happen. For example, setting the stone into a piece of jewelry is always the last thing to do apart from the final polish. Everything else should already be in place. The reason for this is that few stones can withstand heat from soldering or immersion in pickle (see page 60).

Lesson 1: Making accurate scale drawings

Use a sketch (see Unit 3, page 16) as the starting point for a detailed drawing to show the actual size of your piece, where a stone will be set, or features such as surface decoration.

1   Refine a rough sketch by tracing over the most defined outline. Aim for clear, confident lines. Transfer the traced outline onto a sheet of drawing paper, and add a little color (using your preferred medium) to show the type of metal, as well as any stones or other decorations.

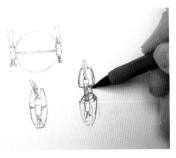

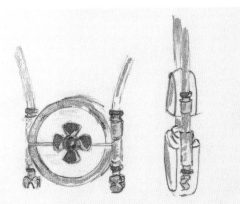

2   Make detailed drawings of how the back and sides of the piece will look, as well as the front. This will help you to decide on the best construction method to use to make the piece. Images should be actual size or drawn to scale.

Lesson 2: Making a working drawing of a ring

The method of construction is often dictated by the design. There may be several parts, each constructed differently. Plan which must be made first and how they will fit together.

1   The ring may be made of separate parts—in this case, the ring comprises four parts. First draw the round shank of the ring to scale.

2   The top is formed from a shaped and textured disk. Make scale drawings of this from above and from the side to show how it curves.

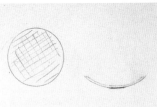

3   Wire decoration is to be added to the shaped disk. Make a drawing from above to show the exact placement of the wire.

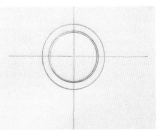

4   A bezel is soldered to the center inside of the disk. Draw this to scale. The series of drawings shown here are very specific—adapt the method to suit each piece you design.

Unit 5: Ordering metals

You can buy metals in many different shapes and sizes. These include sheets, tubes, rods, casting grains, and wire in all kinds of profiles. Metals are usually supplied by a metal dealer, who will need to know the exact dimensions of the material you are ordering. Most dealers will supply materials by mail, but it can be useful to go to a store to see all the options that are available.

Sheet metal can usually be supplied in whatever size is needed. Occasionally, with cheaper metals for example, you will only be able to buy a sheet of a fixed size. You should know the width, length, and thickness of the sheet metal you need before ordering.

When you buy metals you will pay for the weight, expressed in troy ounces or grams. The price of metal fluctuates. Precious metals in particular are subject to price variations according to world markets. This often has a domino effect on the pricing of nonprecious metals as well.

WIRE

SHEET METAL

CHARACTERISTICS OF SHEET METAL

Sheet metal thickness	Characteristics
8-gauge (3 mm)	Rather thick to cut out and not very easy to bend.
10-gauge (2.5 mm)	Can be cut with a coarse saw blade.
12-gauge (2 mm)	Useful size for a substantial piece. Use a coarse saw blade to cut.
13-gauge (1.75 mm)	Similar to 12-gauge but a little easier to bend.
14-gauge (1.5 mm)	Suitable for ring shanks. A size 00 saw blade will cut.
16-gauge (1.25 mm)	Suitable for most work. A size 00 saw blade will cut.
18-gauge (1 mm)	Very useful size for ring shanks, shaping, and hammering.
20-gauge (0.75 mm)	Useful for dapping, bezels, and smaller decorations.
24-gauge (0.5 mm)	Useful for any smaller work. Ideal for bezels.

Lesson 3: Finding dimensions

To work out the amount of sheet metal you need for a piece,
you will first need to draw all the components of your finished
design to scale (see page 19).

1   Draw a square or rectangle
    around the outside of each
    component of the design,
    and measure the sides.
    This will give the dimensions
    of the sheet metal you need.

2   Choose the metal thickness
    required for the component
    (use the table opposite to
    help you decide). Now record
    the dimensions, expressed
    as follows:
    $1\frac{1}{4}$ in x $1\frac{1}{4}$ in x 18-gauge
    (30 mm x 30 mm x 1 mm).

    If gauge thickness is used,
    for both imperial and metric
    it would be 18. Check the
    measurements of your
    materials with a measuring
    tool (see tools, page 26).

ALLOW FOR EFFECTS
The thickness of the metal
used in this ring allowed for
the fact that decorative lines
were applied with a saw. It
was then given a highly
polished finish.

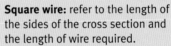

ORDERING WIRE PROFILES

Round wire: refer to the diameter of
the cross section and the length of
wire required.

Square wire: refer to the length of
the sides of the cross section and
the length of wire required.

Oval wire: refer to the height and
width of the cross section and the
length of wire required.

Half-round wire: refer to the height
and width of the cross section and
the length of wire required.

Rectangular wire: refer to the height
and length of the cross section and
the length of wire required.

When ordering wire, ask for:
"Twelve inches of 8-gauge round
wire" or "300 mm of 1-mm-square
wire," for example.

SEE ALSO
Unit 9: Precious metals, page 34
Unit 10: Nonprecious metals, page 38

Unit 6: Testing techniques

Before embarking on a project in precious metal, first try making it in another medium to see how it will work out. It can be costly, and a waste of time, to go straight into metalwork. Make your "mock-up" exactly as you intend the finished piece to be. Any problems or unforeseen techniques will be revealed as the mock-up is being made.

Lesson 4: Making a mock-up

Mock-ups can be made using materials such as paper, thick card, or modeling clay, as well as cheaper metals—copper, copper shim, nickel silver, or lead (see box, right). Try to choose materials that will behave in a similar way to the metal you intend to use. In this example, strips of paper and then copper shim are used to imitate the sheet metal of a finished piece.

LEAD MOCK-UPS

Lead is a very soft metal that is easily worked and therefore good for making mock-ups. But due to its very low melting point, lead should not be used anywhere near where precious metal is worked. If any tiny pieces of lead are left on precious metal when it is being heated, they will melt, leaving holes in the superior metal.

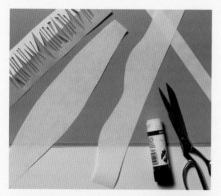

1 Cut a few strips of paper approximately 12 in (300 mm) long and ½–1 in (12–25 mm) wide. Color or mark one side of each strip so that it is different from the other.

2 Try making twists in the strips and gluing or stapling the ends together to form a paper bracelet. Play around with ideas until you find something that you really like.

3 Decide which of the paper mock-ups you like and then cut a strip of copper shim to the same dimensions. Bend it up in the same way as the paper model.

4 You will find that the metal behaves in a different way from the paper. Use your hands to manipulate it to achieve the effect you want. If necessary, use a wooden mallet or metal hammer. Remember, whatever happens in the mock-up will happen when you work on the real thing.

Lesson 5: Making a jig

A jig is used to form metal in a specific and repetitive way. It is worth making a jig if your design requires three or more items of the same shape and size. It may also be necessary to mock-up the jig because it might not be absolutely right the first time.

1 To make a figure-eight chain, you need many identical links. You could form each link individually with a pair of pliers (right), but it will be much simpler and quicker to make the links using a jig.

2 Draw the shape of the link you require onto tracing paper. Use masking tape to fasten the tracing to a block of wood.

3 Find three galvanized nails about as wide as the spaces inside the links, and two slightly narrower ones. Remove the tops with a hacksaw. Sand the tops with 220-grit sandpaper to remove sharp burrs.

4 Hammer the thicker nails into the center of the three areas inside the link on the tracing. Then hammer two smaller nails on either side of the central nail, inside the shape of the link.

5 Take a length of the wire you want to use for the links and hold one end with a pair of pliers. Start to wind the free end of the wire around the nails in a double-ended figure eight, as shown. You may have to anneal the wire first (see page 58) so that it is more flexible around the jig.

6 Repeat the winding process until you have about six complete turns. Lift the wire off the jig and cut through the sides with a saw to make six individual links. If you judge the curve and size to be correct, then the jig is good for repeating the process. If any adjustments need to be made, remove the nails and replace them with different sizes until you achieve the correct link size and shape.

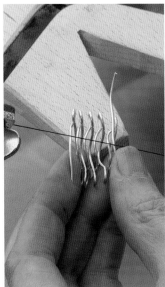

Unit 7: The workbench

A jeweler's bench is the very heart of the workshop. Traditionally, several jewelers sat at one long bench, which had semicircular cutouts in the top surface to provide a workstation for each jeweler. Today, most jewelers work alone. A workbench may be custom built, or you can buy one from any jewelry supply store.

Setting up your bench gives you the chance to arrange everything exactly how you want it. Ease of access to tools should be your priority. Sit down at the bench and try to imagine where each would work best.

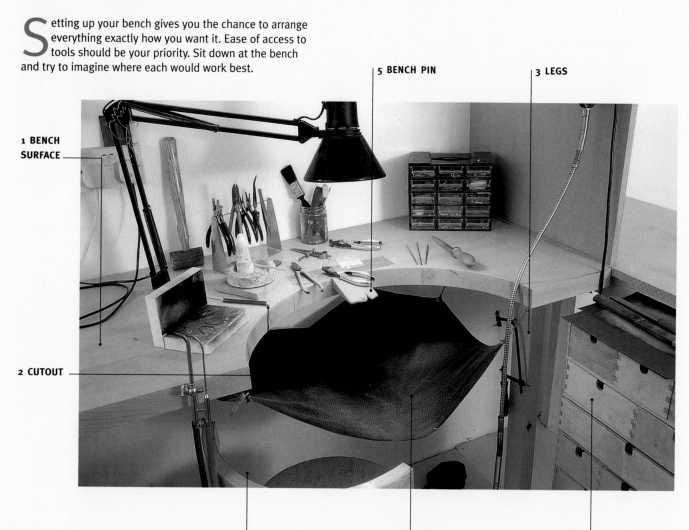

1 BENCH SURFACE

2 CUTOUT

3 LEGS

4 SKIN

5 BENCH PIN

6 DRAWERS

7 CHAIR

1 BENCH SURFACE

The surface of the bench should be 36 in (900 mm) from the floor.
The top surface of the bench should be about 1½ in (30 mm) thick.

2 CUTOUT

The semicircular cutout should have a diameter of approximately 18 in (450 mm).

3 LEGS

The legs of the bench should be sturdy and the whole thing robust enough not to wobble or shake when anything is being hammered on it.

4 SKIN

The leather skin (or sometimes a pullout tray) catches cutoffs and filings. These can be kept and sold back to a metal dealer as scrap.

5 BENCH PIN

The bench pin is like a third hand and is used when filing, piercing, burnishing, drilling, engraving, marking, and for most other work that takes place on the bench. It has one flat side and one sloping side. Usually, the sloping side is uppermost. Some bench pins have a semicircular cutout to support a clamp when setting a stone. It also allows you to brace small items, keeping them steady while you work on them.

6 DRAWERS

It is useful to have drawers or a rack in which to keep tools. Make a home for each of your tools and replace them after every use. They will always be on hand and you won't have to spend hours trying to find them.

7 CHAIR

Your stool or chair should allow you to have your feet flat on the ground if you want, and for you to sit with a straight back and not stoop to see the work. You should be able to see the metal clearly when it is placed on the bench pin. The straighter you can keep your back the better. If you hold your elbows out level with your shoulders they should just sit on the top of the bench. A chair with wheels is useful. Get up every hour or so and have a good stretch.

OTHER EQUIPMENT:

Every bench should have a vise attached to it. Make sure there are some "safe" jaws for use with the vise. You can make these yourself by bending two pieces of 18-gauge (1-mm) copper sheet to fit over the top edges and the insides of the jaws.

Even if you don't have one yet, allow space for a flexshaft motor. It is usually placed just in front or slightly to one side of the semicircular cutout in the bench.

WORKSHOP TIPS

Choose a smooth, plain floor: It is inevitable that you will drop items on the floor. The floor should be smooth (not floorboards with gaps between) and a plain color, so that any speck of metal or tiny stone will be visible if it is dropped.

Wear suitable clothing: Wear a heavy apron in the workshop. Occasionally something hot may fall from the soldering area, and a leather apron will provide good protection against burns. An apron also protects against pickle spillages and dirt from the polisher.

Avoid gas heating: If you need to heat your workshop, choose either electric, solar-powered, or radiant heating. Avoid using a gas heater, which will create condensation and make your tools wet. This will rust them. Gas heaters may also cause headaches.

Look after your tools: The better you look after tools the longer they will last. Don't allow any steel tools to become wet. Keep absorbent paper on hand in the workshop and wipe any tool that you think may be damp before using it. Try not to mark stakes and mandrels by using the hammer at the wrong angle, or they will always leave a mark on any fresh metal that is worked over them. Oil any motorized machine at regular intervals.

Recycle metals: Periodically, put the contents of the skin into a large plastic container. Keep this until it is full. Take this scrap metal to your metal dealer, who will refund you for the scrap value. Keep different metals separate or the value will be less.

Consider the lighting: If possible, your workbench should have natural north light. Bright sunlight shining onto the bench makes it difficult to see clearly. For artificial light, place a light so that it shines over the bench pin.

SEE ALSO
Unit 8: Essential tools, page 26

Unit 8: Essential tools

SEE ALSO
Unit 7: The workbench, page 24

One of the pleasures of jewelry making is collecting the necessary tools. It can take years to acquire everything; good-quality tools are expensive. Some secondhand tools are a great value, but others are not worth considering because they are excessively marked or out of alignment. Avoid buying really cheap new tools. They are rarely worth even the small amount of money they cost because they are unlikely to last.

In this unit the tools are categorized by their uses. Tools marked *** are essential. Those marked ** are desirable but not absolutely essential. Items marked * are often larger pieces of equipment, to which you may have access through a college or other workshop; you may want to consider buying these at a later stage.

Measuring tools

A successful piece often depends on you taking accurate measurements throughout the process.

SCALES*
Choose either small digital scales or the old-fashioned type, with a balance and weights. Use them to weigh metal or to find out the carat weight of stones.

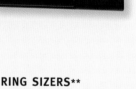

STAINLESS STEEL RULER***
A steel ruler usually comes with both imperial and metric measurements, and it is an invaluable tool.

DIVIDERS***
A pair of stainless steel dividers are used for many types of measurements. They can be used for making the same measurement many times, for example when marking wire for cutting pieces of equal lengths, measuring diameters and lengths accurately, drawing parallel lines on metal, and many other drawing applications.

RING SIZERS**
A ring sizer is a set of rings marked from A–Z that are used to measure fingers if you are making a ring to a specific size (see page 47).

CALIPER**
A caliper is a small sprung instrument that measures anything under 4 in (100 mm). The top "jaws" open out to the width of the piece being measured, and the reading is taken from along the bottom.

RING MANDREL**
A ring mandrel is a tapered, usually aluminum, mandrel, used for measuring ring sizes, with intervals marked A–Z that correspond with the ring sizers. It should not be used as a former.

Cutting and sawing tools

The usual method of cutting silver sheet is with a saw. A guillotine can be used on thicker metals, and snips can be used to cut thin solders. Cut edges may need to be filed smooth or flat.

SAW FRAME***

A saw frame is used with a saw blade for cutting out metal sheet and cutting through wire and tubing. A 6-in (15-cm) wide saw frame* is good to purchase for cutting wider pieces of sheet metal.

SNIPS***

Snips are useful for cutting up solder strips, thin metal sheet, and wire.

SAW BLADES***

Saw blades come in different sizes, ranging from size 5, for cutting very thick metal, down through oo to 8/o, for very fine cutting. Use them with the saw frame.

COPING SAW**

A coping saw is larger than a piercing saw and is used for cutting plastics and wood.

GUILLOTINE*

A guillotine is a heavy-duty piece of equipment. It is useful if you are cutting large pieces of metal into strips but is otherwise not really necessary.

TOP CUTTERS***

The cutting action is on the top of top cutters. They are used for cutting up very small pieces of silver, copper, gold, platinum, and binding wire. They should never be used for cutting stainless steel.

SIDE CUTTERS**

The cutting action of side cutters is on the side. This is not quite as useful as the top cutters, but they can be used in a similar way.

TUBE CUTTER**

A tube cutter is a small handheld tool used to hold tubing, with an adjustment facility that allows you to cut off pieces of equal lengths.

Bending tools

Pliers come in a range of shapes for different kinds of bending applications, from making smooth curves to forming jump rings.

PARALLEL PLIERS***

Parallel pliers are used to straighten out metal sheet and thick wire, for holding pieces of metal level for filing, and for closing thick rings. They come in flat and round versions. A third version has a plastic interior covering over the jaw, to protect the sheet or wire.

TAPERED FLAT PLIERS**

Tapered flat pliers have a flat inside face and taper to a blunt pointed end. They are used to close awkward or small jump rings and hold other small items.

ROUND-NOSE PLIERS**

Round-nose pliers are used for making individual circles or jump rings. The wire is held between the two ends and wrapped all the way around, then cut through to make the circle.

FLAT-NOSE PLIERS***

Flat-nose pliers come in a range of sizes and are used to bend sharp corners in wire and metal sheet, and for holding things flat, straightening wire, and closing jump rings.

HALF-ROUND AND ROUND/FLAT PLIERS***

Half-round and round flat pliers are used for bending wire and metal sheet into a circle without leaving marks. The flat side is held against the outside of the curve, and the round side is used to make the curve on the inside.

Holding tools

Use vises and clamps to hold pieces firmly while you work them.

BENCH VISE***

There are two types of bench vises. One is quite small, able to turn in all directions, and has "safe" or plastic jaws. The other is a more heavy-duty vise, which is used to hold stakes, mandrels, and draw plates. Both types should be fixed permanently to the bench.

RING CLAMPS**

Wooden handheld clamps with leather pads are used to hold rings safely, without damaging their shanks, for stone setting.

Files

You will need a selection of files in different sizes for smoothing cut edges and shaping curves.

FILES AND NEEDLEFILES***

Files vary in quality; get the best that you can afford.

Flat file—for flat surfaces, filing between joins to be soldered, edges, and outside curves;

Half-round file—for inside curves and edges;

Triangular file—for filing around the top edges of bezels, grooves for right angles, and other difficult edges;

Square file—for making right angles true and filing inside areas;

Knife—with one thick edge and one thinner edge, it is used for getting in between small areas.

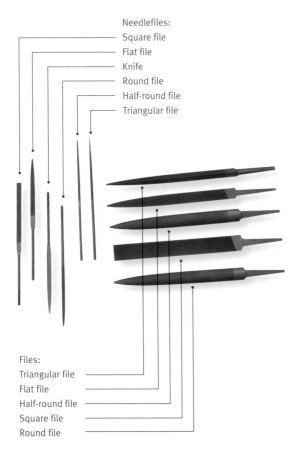

Needlefiles:
Square file
Flat file
Knife
Round file
Half-round file
Triangular file

Files:
Triangular file
Flat file
Half-round file
Square file
Round file

Cleaning tools

Metal needs cleaning after soldering and after a piece has been completed. It may also need cleaning after it has been worn several times. Only use abrasive cleaners before the work is complete because they will scratch and dull a polished surface.

ULTRASONIC CLEANER*

An ultrasonic cleaner is a plastic or steel container with a basket hanging inside it. It is used with articles that have just been polished and have remains of black polish on them. The container is filled with water, liquid soap, and household ammonia. Articles are placed in the basket and ultrasonic rays pass through the liquid, which removes the greasy polish. The liquid works better if it is hot.

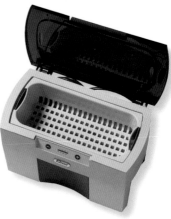

PUMICE POWDER**

Pumice powder is a fine gray powder. It is mixed with water to make a paste that is then rubbed onto metal to clean it after annealing or soldering. Apply it to a wet green scrubbie or Scotch-Brite pad and it will clean metal quickly. Rinse thoroughly.

SILVER CLOTH***

A silver cloth is a duster that has been impregnated with a metal cleaner. It can be fastened into the bench vise and pulled tight while the article is rubbed onto it.

LIQUID SOAP***

Liquid soap can be used neat in the palm of the hand with some running water to help clean an article that has polish grease on it.

Forming tools

Forming tools are made from steel or wood and are used to form metal around. The metal is held against the former and hit with a hammer or mallet so that it gradually takes on the shape of the former.

STAKES*

Stakes are shaped metal formers. They are used mainly by silversmiths, to raise and forge metal sheet when making larger pieces and vessels.

SWAGE BLOCK*

A swage block is a metal block with semicircular lengths molded through both sides. The handles from wooden or metal dapping punches are used to shape the metal.

MANDREL**

A mandrel is a tapered steel or wood former that is used to shape metal. Mandrels can be round, oval, square, teardrop, or hexagonal in cross-section; the one shown here is a round ring mandrel.

DAPPING BLOCK**

A dapping block is a brass or metal cube with different size half-spheres molded into each side. It is used to form round metal disks into domes.

BEZEL-FORMING PUNCHES AND BLOCKS*

These are similar to dapping blocks and punches, but are used to form a cone shape, for setting faceted stones. They can be round, oval, rectangular, square, hexagonal, and other shapes.

DAPPING PUNCHES**

Dapping punches are shaped to fit into each different size of half-sphere in the dapping block. They can be made of wood or steel, and are placed on top of the metal disk and hit with a hammer or mallet to form the dome.

Hammering tools

Hammers come in a range of sizes and materials. Unless you are doing a lot of silversmithing work, start with a small selection.

RIVETING HAMMER***

A lightweight riveting, or jeweler's, hammer can be used for all delicate work. It has one flat end, which can be used for riveting and other small jobs, and one wedge-shaped end which can be used for texturing metal.

BALL-PEEN HAMMER ***

A ball-peen hammer has a metal head with one flat end and one rounded end. The round end is used to shape and texture metal, and hammer in small spaces. The flat end is used for stretching metal on a stake or mandrel, and tapping the end of punches and repoussé tools. Choose a medium-weight hammer.

LARGE, HEAVY HAMMER***

A large, heavy hammer can be used for all heavy work. It can forge out lumps of molten silver, give a heavy texture on metal, and be used for reshaping.

MALLET***

A leather rawhide, wooden, or plastic mallet is used to hammer metal without leaving a mark. A mallet is usually used to shape metal without stretching it.

Drilling tools

Holes in metal can be drilled with an electric drill or by hand. If the metal is less than 18-gauge (1 mm) thick it is easier to drill by hand. Before drilling any hole, mark its position with a sharp, pointed punch, so that the drill bit does not "wander."

PUNCH**

A punch is a little metal tool, similar in shape and size to a pencil. One end has a point and the other is blunt and flat. The point is used to make a mark in the metal where a hole is to be drilled. The blunt end is struck with a hammer.

ARCHIMEDIAN DRILL**

An archimedean drill is a handheld push-action drill which allows the work to be held in one hand while the drill is pushed up and down with the other.

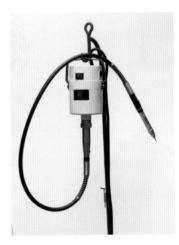

FLEXSHAFT MOTOR**

A flexshaft motor is attached above the workbench, and has a flexible driveshaft. Anything from small drills, abrasive cutters, grindstones, polishing wheels, felt brushes, and muslin mops can be fixed into different sized collets or mandrels, making it a very useful addition to your tool collection.

HAND DRILL***

This is a small, handheld drill which fits into the palm of your hand. Hand drills can hold different size chucks, to allow for many different size drill bits, from 76-gauge (0.5 mm) up to about 32-gauge (3 mm).

Soldering tools

Many pieces will need to be soldered (see page 72). Soldering involves flames and very hot pieces of metal. If possible, the soldering area should be separate from the workbench area. If you do have your soldering block on the bench, make sure there is absolutely nothing that will get in the way of the flame.

SOLDERING STAND***

The base for your soldering area should be a steel plate of some sort. This could be a specially made piece of stainless steel or an old ovenproof pan. On the steel base, place a revolving soldering stand. This allows you to turn the work around slowly as you solder, making it easy to see the join as the solder flows and to check whether it has all run.

TORCH***

A torch is used to heat the metal to annealing temperature, for all soldering, or for melting metal for casting. There are several different types of torches.

QUENCHING BOWL***

Place a toughened glass bowl of water close to the soldering area. Use a pair of insulated tweezers to "quench" the work in this water after it has been soldered or annealed (see page 60).

INSULATED TWEEZERS***

Insulated tweezers are sprung tweezers with insulation on the handles designed to withstand heat. They should be squeezed together to open them. They have either straight or curved ends, which makes them useful for holding pieces together when soldering. They should never be placed in acids or pickles.

SOLDERING BLOCK OR FIREBRICK***

On top of the soldering stand, place a soldering block or firebrick, or a charcoal block. A soldering block will not deteriorate as quickly as a charcoal block, but there is more reflective heat from the charcoal. Use charcoal very sparingly, to minimize any negative impact on the environment.

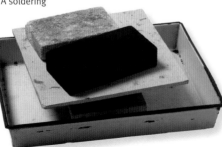

SOLDERING WIG***

A soldering wig looks a little like a wire bird's nest. Place articles to be soldered on the wig, which will help to distribute heat around them evenly. The wires can be moved to form different supports.

BORAX AND DISH***

The borax cone is dipped in water and rubbed around the dish to make a whitish flux paste. The paste or flux is placed in and around a join before soldering. Borax is one of several available fluxes.

TITANIUM PICK**

A titanium pick is used to encourage solder to flow around a join by drawing it along as the solder is flowing. It can also be used to place solder paillons onto the work as it is being heated up.

BINDING WIRE***

Steel binding wire comes on a reel and is supplied in light, medium, and heavy weights. It is used to hold pieces together when soldering and for measuring ring and bracelet lengths. It should always be removed before the article is placed in the pickle.

Polishing tools

Polishing can be done by hand or with an electric motor. Hand polishing needs only a selection of wet and dry papers, fine finishing cloth, and an impregnated silver cloth. A polishing machine requires a variety of mops charged with different polishes.

POLISHING WHEEL*

A polishing wheel is an electric motor with a horizontal shaft. Polishing mops and brushes can be screwed onto the shaft and polish or liquid soap applied to them. There should be some means of extraction in the workspace because the polish can be rather dirty and create a lot of dust.

BUFFING WHEEL*

The round buffing wheel, which is made from calico, can be different widths. It is screwed onto the shaft of the polishing wheel and charged with the appropriate polish.

POLISH BARS*

Tripoli: a brown bar, used at the first polish

Green: a green bar, used as a general-purpose polish

White diamond: a white bar, used to give a high polish

Rouge: used for the final polish

STAINLESS STEEL WHEEL*

A stainless steel wheel has lots of fine blunt steel pins that give a satin look to a metal surface. No polish is required with this wheel.

LAMBSWOOL BUFFING WHEEL*

A lambswool or muslin buffing wheel is a very soft mop used with the last coat of "rouge" polish.

WET AND DRY PAPERS***

Wet and dry papers come in varying degrees of abrasiveness, from very rough to very fine.

BURNISHER***

A burnisher is a polished stainless steel tool that is rubbed firmly against metal edges to give a high shine.

Unit 9: Precious metals

Platinum, gold, and silver are precious metals. In their pure state they do not oxidize and are not corroded by acids. If you have ever seen an ancient gold coin that has just been dug up from the ground, you will appreciate how the dirt just rubs off to reveal the gold in exactly the same state as it was when it was minted all those centuries ago.

SUBTLE FINISH
This pendant contrasts silver and 18-carat gold given the same matte finish with wet and dry papers followed by pumice powder.

It is mainly the precious metals that are used for jewelry making. They have properties that make them ideal for bending, shaping, and polishing and for being able to show off, to their best advantage, any stones or enamels that are set into them.

Unfortunately, gold and silver are too soft in their pure state to be practical for most jewelry applications, so they are combined with other metals to make alloys, to make them suitable for different uses, and in a range of colors and costs. These alloys are still known as precious metals, but their quality is strictly controled and guaranteed by the hallmarking process.

To start out, and for making mock-ups, metals such as copper, nickel, and silver can be used. Stainless steel, which is very hard, is used for pressing into precious metals to get a textured finish, but it is not practical for jewelry making. As you become more confident working with metals, you may want to work with silver, before moving on to the different carat golds and to platinum.

Properties of precious metals

Each metal and alloy has a unique melting temperature and specific gravity (density). These properties are useful to know when you are working with metals.

PLATINUM (Pt)
Platinum is a dense white metal that is often used for mounting diamonds. It is the most expensive of the noble metals. Owing to the popularity of white metals in recent years, platinum has become hugely popular. It is hardwearing and relatively easy to work but must be soldered at very high temperatures.

Because it is so pure, platinum does not oxidize when heated, so solder joins do not have to be fluxed and consequently, platinum work does not need pickling. Platinum can be given a highly polished finish before soldering that will not be affected by the heat. If platinum is heated incorrectly, with a reducing flame, it can become grainy and brittle.

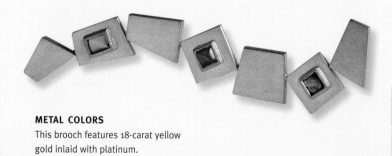

METAL COLORS
This brooch features 18-carat yellow gold inlaid with platinum.

	Melting temperature	Specific gravity
Platinum	3217°F (1769°C)	21.45

	Gold:alloy ratio	Melting temperature	Specific gravity
24-carat gold	24:0	1945°F (1063°C)	19.32
22-carat gold	22:2	1724°F (940°C)	17.2
18-carat gold	18:6	1660°F (904°C)	16.15
14-carat gold	14:10	1476°F (802°C)	13.4
10-carat gold	10:14	1447°F (786°C)	11.6
9-carat gold	9:15	1616–1652°F (880–900°C)	11.2

COLOR VARIATION
These metals in sheet form demonstrate the differences in color that occur in different types.

GOLD (Au)

Gold is described in carats. The word "carat" is also used to describe the weight of a stone, but when referring to gold, it describes the proportion of gold to alloy that makes up 24 parts.

Silver, palladium, copper, and occasionally zinc are added in different proportions to make different colored golds. The lower the carat the more different colors can be created because there are more parts that are not gold. Melting temperatures vary in different alloys; for example, 18-carat white gold has a higher melting point than 18-carat yellow gold. White gold contains palladium, which has a higher melting point than the silver and copper in yellow gold.

Like all precious metals, gold should be annealed before it is worked (see page 37). Higher carat golds stay softer for longer than those below 14 carat because of their higher gold content. Metal will only harden as it is worked, so if annealed gold is left for several days, it will not have hardened by the time you resume work on it.

SEE ALSO
Unit 10: Nonprecious metals, page 38
Unit 15: Annealing, page 58
Unit 21: Soldering, page 72

SILVER (Ag)

Pure silver 999.9 is a bright white metal that is too soft for most jewelry-making applications, with the exceptions of enamelwork, granulation, and some chainmaking. For this reason, a small amount of copper is added to pure silver to make a more durable metal—known as standard or sterling silver. Out of 1,000 parts, 925 are silver and the remaining 75 are copper.

Other silver alloys have a lower silver content, but these do not qualify for the legal marking of 925. Pure silver does not oxidize. Because of the copper content in 925, it oxidizes when heated and if left exposed to the atmosphere. Silver is a malleable metal but will workharden after a while. It is annealed to keep it soft and workable.

	Melting temperature	Specific gravity
999.9 silver	1760.9°F (960.5°C)	10.5
925 silver	1640°F (893°C)	10.4

Properties of silver and gold solders

Different solders come with a range of properties and flow temperatures. Always use solder that is appropriate for the type of metal and the task.

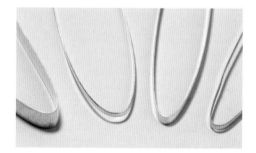

SILVER SOLDERS

Type	Use	Flow temperature
IT/enameling	Use only if piece is being enameled.	1488.2°F (809°C)
Hard	Always use first.	1448.6°F (787°C)
Medium	Use at intermediate stage. Can be sticky.	1358.6°F (737°C)
Easy	Excellent for final solders.	1324.4°F (718°C)
Extra easy	Use as a last resort. When used on silver it tarnishes quickly.	1140.8°F (616°C)

GOLD SOLDERS

Type	Use	Flow temperature
22-carat	Use only with 22-carat gold.	There is only one solder for 22-carat gold. It melts between 1589–1652°F (865–900°C).
18-carat	Use only with 18-carat gold.	Hard: 1455–1525°F (790–830°C) Medium: 1350–1410°F (730–765°C) Easy: 1290–1320°F (700–715°C)
9-carat	Use only with 9-carat gold.	Hard: 1390–1465°F (755–795°C) Medium: 1355–1390°F (735–755°C) Easy: 1200–1330°F (650–735°C)

GOLD SOLDERS
Gold solders are available in yellow, white, or red. They come by the pennyweight (dwt) or in rectangles approximately 7/8 in x 6/8 in (20 mm x 15 mm). Instead of being marked hard, medium, or easy, white-gold solder is marked numerically. It is supplied in carats and should be used with the appropriate carat of gold.

SILVER SOLDERS
Silver solders come in long strips of different thicknesses and widths.

SEE ALSO
Unit 19: Cutting, page 68
Unit 21: Soldering, page 72

Lesson 6: Annealing gold

Gold alloys are annealed in a different way from other metals. (For more on annealing silver, see Unit 15, page 59.) Some are quenched and pickled directly after annealing, others are left to cool in the air before pickling, and some are allowed to cool down to "black heat," at around 1040°F (560°C) before pickling. Usually your supplier will tell you how to anneal your gold, but if you are unsure, use the following method.

1   Cut a strip of 9- or 18- (or both) carat white, yellow, or red gold. Place it on your soldering block and heat it with a soft nonoxidizing (reducing) flame.

4   Anneal the piece again, and this time leave it on the block to cool completely. Then pickle to clean it and try bending it again. If it is still hard, go to step 5.

2   Place the gold into water to quench it; then clean it in the pickle.

5   Anneal the gold again and leave it for about half a minute before quenching and pickling.

3   Test the gold by starting to bend it with a pair of pliers to see if it is soft. If it bends easily it has been annealed correctly. If it is hard to bend, go to step 4.

6   Try bending it again with the pliers. Make a note of how this particular piece of gold was annealed.

Unit 10: Nonprecious metals

Nonprecious metals such as aluminum, lead, zinc, titanium, and niobium may be used for jewelry making, often for the colors that can be achieved through anodizing or heating them. Nonprecious metals should be kept well away from precious metals to avoid contamination as heat is applied.

Properties of nonprecious metals

As with the precious metals, each metal and alloy has its own unique properties. These are useful to know when you are working with metals.

LEAD (Pb)

Lead is a pure, soft gray metal. It is usually used as a support when working on other metals or as a model.

	Melting temperature	Specific gravity
Lead	621°F (327°C)	11.4

ZINC (Zn)

Zinc is a pure white metal, usually added to other metals to make an alloy. It is used in silver solder because it has a very low melting point.

	Melting temperature	Specific gravity
Zinc	787°F (419°C)	7.1

ALUMINUM (Al)

Aluminum is a gray, slightly grainy, light metal that can be turned on the lathe easily but is difficult to shape and cannot be soldered. Its main use in jewelry is when it is anodized to give a variety of colors.

	Melting temperature	Specific gravity
Aluminum	660°F (122°C)	2.7

TITANIUM (Ti)

Titanium is a pure, hard (but light) white metal with a very high melting point, which makes it impractical to solder. It can be anodized and used to make large but very light pieces of jewelry.

	Melting temperature	Specific gravity
Titanium	3272°F (1800°C)	4.5

COPPER (Cu)

Copper is a pure, soft brown metal that is quickly workhardened. It is alloyed with silver and gold to make them more practical or to affect their color. When you anneal copper, it will first turn black, then will turn pinkish. When you see this pink color, quench the metal.

	Melting temperature	Specific gravity
Copper	1981°F (1083°C)	8.9

Lesson 7: Workhardening metal

A metal's molecular structure is compressed during hammering and shaping; it becomes workhardened and will need annealing before it can be worked again.

SEE ALSO
Unit 15: Annealing, page 58
Unit 21: Soldering, page 72
Unit 24: Polishing and finishing, page 84

1   Take a small piece of copper sheet approximately 3 in x 1 in x 18-gauge (80 x 20 x 1 mm). Place it on a soldering block and heat it with a torch until it goes completely black.

2   Pick the copper up with a pair of insulated tweezers.

3   Drop the copper in a bowl of cold water. Watch how the black seems to flake off. The black coating is copper oxide (known as firescale).

4   Pick the copper out of the water with a pair of stainless steel tweezers and place it into warm pickle (see page 60). Leave it for a few minutes and then remove it with the steel tweezers and rinse it in cold water. It should be cleared of the firescale (see page 84 for more on firescale).

5   Try bending the copper strip between your fingers and thumb. It should be very malleable. Straighten it out again and place it on a solid surface, either steel or wood, and hit the metal all over with the rounded end of a ball-peen hammer.

6   Now try bending the strip again between your fingers and thumb. You will find it is much harder than the first time you bent it. The copper is "workhardened."

USING COPPER
Copper is a reddish brown metal and comes in sheet and wire form. When annealed, it is very soft and an ideal metal for making mock-ups. It is soldered with silver solder.

Unit 11: Stones and beads

At some point you will want to incorporate stones or beads into your work. Stones add color, depth, and focus to a piece of jewelry. There are many ways of setting stones into metal, and a few of the basic techniques for doing so appear in Unit 32, page 108. The intention of any setting technique is to show a stone to its best advantage. It may be a very simple rub-over setting or a detailed claw setting, but in all cases a setting is made to fit an individual stone.

Stones are available in a huge variety of colors and sizes, and they can range from being relatively affordable to priceless. The appearance of any stone depends on how it has been cut to reflect light. On these pages you can see a variety of precious and semiprecious gemstones in a range of different cuts.

STONE SHAPES

Cabochon: Cabochon stones usually have a flat—or nearly flat—base and a smooth domed top. They may be any shape, but oval or round cabochons are most common. To hold them into a setting, a small collar, or "bezel," is made that fits neatly around the outside edge of the stone and is gradually coaxed over and down onto it.

Faceted: Faceted stones have many flat angled surfaces that combine to reflect light in an eye-catching way. Precious stones, such as diamonds, sapphires, emeralds, rubies, spinels, and topaz are usually given a faceted cut. They can be all sorts of different shapes, but the cut most often given to any of the shapes is "standard brilliant cut."

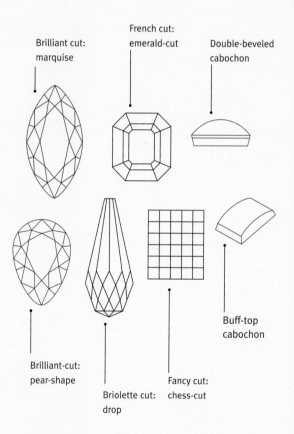

Brilliant cut: marquise

French cut: emerald-cut

Double-beveled cabochon

Brilliant-cut: pear-shape

Briolette cut: drop

Fancy cut: chess-cut

Buff-top cabochon

CUTS OF STONES

Cuts vary, depending on the type of stone and the desired effect. Variations on the brilliant cut, for example, are designed to let light reflect through the stone. Fancy cuts are used to retain weight or sometimes to disguise flaws.

BUYING STONES

There is such a fantastic array of beautiful stones and beads available that it can be difficult to know where and what to buy. Many large towns and cities have a dedicated bead store where you will find individual beads and different strings of semiprecious beads that can be simply made up into necklaces. There is also a growing number of gem shows where individual stone importers and dealers show their wares, and these can be found in many different locations around the country. As well as a massive array of strung beads to choose from, you will find some lovely cabochon and faceted stones. Internet sales are amazingly popular, but at first, it is advisable to buy just one stone or a single string of beads to minimize risk until you find a dealer you can trust.

CARATS

Stones are usually sold by weight, which is measured in carats. (You may see it spelled with a "k" rather than a "c".) There are 100 points in a carat, so if you know that a stone is a "10-pointer," that means it weighs one tenth of a carat.

To determine the carat weight of a stone:

• Weigh the stone in grams

• Multiply the weight in grams by five; this is the number of carats

To find the cost of the stone, multiply the number of carats by the cost per carat of the stone. Some stones are sold per item, but this usually applies to cheaper stones such as turquoise, quartz, and hematite.

QUALITY

Generally speaking, the clearer the stone, the better the quality. However, remember that very clear flawless stones may have been manufactured and heat-treated—which is fine, as long as you know that is what they are. More expensive stones such as rubies and emeralds may have been enhanced by heat, treated with lasers, or injected to give them a better color. Unless stones are very expensive they are unlikely to be totally unflawed; and if the color looks too good to be true it probably is. Most stone suppliers have a showroom, so contact them first and ask to go and see what they have in stock. Pick out one or two things that you really like and can afford, and don't feel pushed into buying more than you want.

OPALS

Opals sometimes have a dark backing applied to them, to enhance the colors of the stone; these are called doublets or triplets. Opals without any backing usually look quite milky (unless they are orange fire opals or black opals).

TURQUOISE

Turquoise are very soft; they break easily if there is too much pressure put on them while setting or stringing. The most valuable turquoise are the rather muddy green ones, with black lines running randomly through them (the matrix).

SEE ALSO
Unit 27: Fittings, page 92
Unit 32: Stone setting, page 108

STARS AND CAT'S-EYES

Stars generally appear in sapphires and rubies. If they are held in the light you should see strong thin lines radiating out from the center in the shape of a star. If the stone is very clear as well as having a star, it will be expensive. If the stone is rather opaque with a star, it is usually less expensive. Cat's-eyes appear as a thin line right across the stone and seem to move with the light. They add value to it. Both cat's-eyes and stars only appear in cabochon stones.

This sapphire has a cat's eye.

TOURMALINES

Tourmalines are very popular stones. They come in a fantastic range of colors, from blue, to turquoise blue, gray, dark green, light green, pink, and yellowish. Sometimes two or three of those colors appear in the same stone. A clear dark green or pink stone can be very expensive.

STONE-BUYING TIPS

Look out for large inclusions that cross a stone from side to side. They may appear very fine but could cause the stone to split if too much pressure is applied during setting. Flaws in faceted stones are less obvious. If you are buying an expensive stone, have a look at it through a x10 magnification loupe, because any flaws will show up then.

Diamonds have a universally understood grading of quality. They range from "flawless" to "imperfect," and the price per carat depends on the grading. When choosing or ordering diamonds, ask which grade you are looking at.

FL—Flawless
Perfect. No internal or external imperfections can be seen with the eye or a x10 loupe.

IF—Internally flawless
No imperfections or inclusions can be seen under a x10 magnification loupe.

VVS—Very very slightly imperfect
May have very small inclusions that can only just be seen under a x10 loupe.

VS—Very slightly imperfect
Small inclusions, quite difficult to see under a x10 loupe.

SI—Slightly imperfect
Inclusions are quite easily seen under a x10 loupe and a trained eye may spot them without a loupe. Another potential flaw in diamonds is the appearance of little black specks of carbon.

I—Imperfect
Inclusions that may be seen with a naked eye.

There are grades within these ranges that denote finer and more specified gradings, but these grades should give a useful starting indication.

DIAMOND SETTING
A pair of drop earrings showing pearls combined with diamonds.

Lesson 8: Stringing a bead necklace

Some semiprecious stones are supplied with drilled holes, to use as beads. You can also buy a wonderful selection of beads made of all sorts of other materials. Thread beads onto earring drops or pendants, or thread them to make a simple necklace, as shown here.

NECKLACE CATCH
Allow 1 in (25 mm) for the catch and hook of a bead necklace. This would mean that you would only need to string 17 in (420 mm) of an 18-in (450-mm) necklace.

1   Necklaces are usually 16–20 in (400–520 mm) long. Cut a length of tiger tail or nylon thread at least 6 in (150 mm) longer than you need it.

2   Thread an end-clasp onto a jump ring and solder the jump ring closed (see page 74).

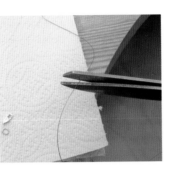

3   Thread the tiger tail or nylon thread through the jump ring and bring it back on itself.

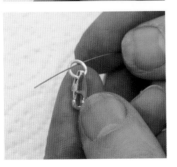

4   Thread a crimp onto the other end of the thread and bring it close to the jump ring. Thread the second end through it. Squeeze the crimp tight with the tapered flat pliers.

5   Now thread the beads onto the thread. Push them right up to the crimp and thread the short end through as many of the beads as possible. Trim it if necessary.

6   When you have finished threading all the beads, thread on the second crimp, then the fastening jump ring with the end catch attached. Pull the thread tight.

7   Thread the end through the crimp. Push the crimp right up to the jump ring, and squeeze with the tapered flat pliers. Thread the short end of the thread back through as many beads as possible.

chapter 2
techniques

With lots of ideas bursting to become reality, you will be ready to use the practical part of the book. One important thing to remember is that jewelers do things in a variety of different ways. The techniques explained in the lessons show one way of working, but there is often more than one way to achieve a similar effect; you may find your own methods and ways of making a technique work better for you. This section includes enough information to help you through the many basic skills, and to encourage you to pursue—and, most important, to enjoy—your jewelry making.

Unit 12: Measuring

Jewelry is designed to be worn—an aspect of design that can get overlooked. Finding the correct weight, sizings, lengths, balance, and placement of fittings, which all ensure that jewelry is a pleasure to wear, are all dependent on good measuring in the first instance.

After deciding on a design, the measurements are the next thing to work out. Sometimes, you will be able to work them out by eye. At other times, accurate drawings will be required (see page 19). Certain formulas are useful for calculating the lengths needed for ring shanks and bezels, and for making different geometric shapes, as well as being able to calculate the cost by weight of a piece of metal or a stone.

Many of these calculations involve pi (π), which is expressed either as 3.142, or as 22 divided by 7. Let pi become your friend, and it will do all your calculations for you!

RING MEASUREMENTS
Even a simple ring with a rub-over setting requires accurate measurements. It is important to know the wearer's ring size to calculate the length of metal for the shank, and the size of the stone to work out the length of sheet metal needed for the bezel.

WORKING OUT WEIGHT

Scales are useful for weighing metal and stones. You can work out the weight of an object—a piece of metal for example—without scales, but you first need to know its volume and density.

Volume:
Sheet metal = length x width x thickness
Round wire = area of face (π x radius^2) x height (or length of wire)

Density:
Each metal has its own density, or specific gravity (see pages 34–35).

Weight:
Weight = volume x density
For example, to find the weight of a piece of 18-carat gold that measures 20 mm x 60 mm x 1 mm:
20 x 60 x 1 = 1200
1200 x density (16.15) = 193.8
so it weighs 19.38 g.

USEFUL FORMULAS

Circumference of a circle:
diameter x π
(diameter x 3.142)
You will need to know the circumference to find the length of a ring, the size of a bezel to fit around a stone, or for any other circle you are making.

Circumference of an oval:
$$\left(\frac{\text{length} + \text{width}}{2} \right) \times \pi$$
You will need to know the circumference of an oval when making bezels for oval stones or for any other oval you are making.

Area of a circle:
πr^2
(π x radius squared)
You will need to know this when working out the weight of a circle or a round wire section.

Area of a triangle:
base x ½ height
You will need to know this to work out the weight of a triangle or a triangular wire section.

Volume of a sphere:
$$\frac{4\pi r^3}{3}$$
(4 x π x radius cubed, divided by 3)
You will need to know this if you are making solid balls.

Carat weight (of a stone):
weight of stone (in grams) x 5
You will need to know this for all sorts of applications.

Lesson 9: Finding a ring measurement

To make a ring that fits, you must cut a length of metal accurately. To find the length of metal required you need to know what size you want the ring to be. There are two ways of finding the length of metal to cut.

1   For the first method, simply measure the diameter of the ring size needed (use a ring sizer) and multiply the answer by pi.

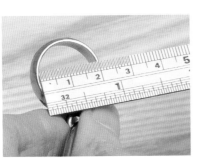

2   For the second method, tightly fasten a piece of binding wire around a ring mandrel on the marking for the size required.

3   Lift the binding wire off the stick, cut it open, and spread it out into a straight line (which can then be measured easily).

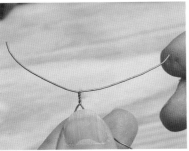

4   For both methods, measure the thickness of the metal to be used with a pair of calipers and double it. Add that measurement onto the length of the ring shank.

5   Once you have found the required length for the band, you can scribe it onto your sheet of metal ready for piercing (see Unit 12, page 49).

RING SIZES

Ring sizes in the United States are measured on a numerical scale. A normal size range would be between 4 and 15. There are different standard ring-sizing systems in different parts of the world. (See page 141 for an international ring-size conversion chart.)

PRACTICE PROJECTS

Use this technique to make the cabochon-set ring; see page 126.

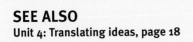

SEE ALSO
Unit 4: Translating ideas, page 18

Lesson 10: Finding a bracelet measurement

A closed bracelet has a different measurement from an open one. Most closed bracelets measure between 8 and 9 inches (200–230 mm), and open ones between 6½ and 8 inches (170–200 mm). It is easiest to use a bracelet measurer to help you find the measurement, but you could simply use a tape measure.

1 Open the bracelet measurer and slip it over the widest part of the hand, without pushing too hard. If you are making a closed bracelet, the correct measurement will be where it fits at the widest point.

2 If you are making an open bracelet, the correct measurement will be where it fits neatly on the wrist.

OPEN BRACELET
The curves and lines of this open bracelet were influenced by the original ideas from nature on page 17.

CLOSED BRACELET
The loose rings made from silver and wood on this closed bracelet would have been fitted after the shape was formed.

PRACTICE PROJECTS

Use this technique to make the etched bangle; see page 130.

Lesson 11: Measuring lengths

Use this method to cut several lengths of wire,
each of which is exactly the same length.

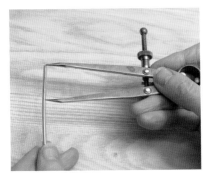

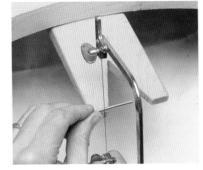

1   Work out how long you want the lengths of wire to be. Set your dividers to the length required.

2   File one end of the wire flat. Hold one end of the dividers against the flat edge and use the other end to scribe a line around the wire.

3   Use the saw to cut where the line is scribed. File the end flat again and mark the next cutting line with the dividers.

4   You can use the same method for scribing the width of equal-size bands to be pierced from a sheet of metal.

MAKING LINKS
It is useful to be able to make many pieces of wire of exactly the same length, to form uniform links in a bracelet or necklace for example.

Unit 13: Transferring patterns

After a design for a cutout piece is finalized on paper, it must be transferred onto the metal for piercing. Often the metal is flat, and the pattern can be transferred easily. There may be times when the metal is shaped first—into a dome, for example—and the pattern is cut afterward.

The line to be cut should be distinct. It is never satisfactory to cut to a line that is feathery or has one or two other lines near it. This is particularly true when cutting circles (for more on piercing, see page 52). There are several methods for transferring the design to the surface of the metal—choose the one that is most appropriate for your design. Whether the line is drawn onto tracing paper that is stuck onto the metal or scribed directly onto the metal, it should be clear and easy to follow.

Lesson 12: Fixing a pattern onto metal

Marking out the pattern on tracing paper and then using the saw to cut along the lines of the pattern is the simplest way of cutting out. Make sure the tracing is stuck down firmly onto the metal, and blow away the dust created as you are sawing.

1 Trace your finished design onto tracing paper with a fine mechanical pencil. Keep the outline clear, with no sketchy lines.

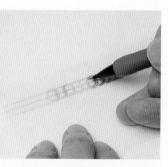

2 Cut out the tracing paper, leaving a margin around the outside of the pattern.

3 Remove any protective plastic from the topside of the metal sheet.

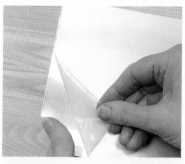

4 Use wet and dry paper to remove scratches on the metal surface before cutting. Apply glue to the back of the tracing paper and allow it to air-dry. Place it down onto the metal, making sure that the traced cutting line is not over the edge of the metal. Cut out the pattern with the saw.

Lesson 13: Tracing the pattern onto the metal

This method is an alternative to using tracing paper.

1 Clean the metal with a piece of wet and dry paper. Use it dry and begin with a fairly coarse one (240–400 grade), and then go over it again with a 400–600 grade.

2 Trace the shape onto a sheet of tracing paper. Turn the tracing over and trace over the line with the pencil, pressing down hard to leave a good mark.

3 Secure the metal on a flat surface with masking tape. Fasten the tracing (right-side up) on top of the metal with masking tape. Trace along the line with the end of a scribe, pushing down firmly, so that the pattern is transferred onto the metal.

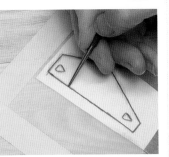

4 Remove the tracing and carefully go over the lead pencil line with a metal point. Cut out the pattern with the saw.

Lesson 14: Drawing the pattern onto the metal

You can draw your pattern directly onto the metal. Use this method if the pattern needs to be extremely accurate, such as a cone to make a setting for a faceted stone (see page 114).

1 Work out on paper the exact pattern for the cone (or whatever you need to cut) accurately. For a cone this is usually done with a pair of dividers or a pair of compasses, and has a center point.

2 With masking tape, fix down the metal close to your drawing, so that you can take all your measurements easily from the drawing. Start to mark the pattern onto the metal.

3 The line is now ready to be cut along with the saw. If you use dividers, remember they will scratch the surface. Only scratch in areas to be cut, because the scratches can be difficult to remove.

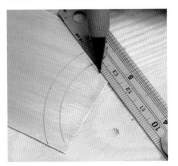

SEE ALSO
Unit 9: Precious metals, page 34
Unit 10: Nonprecious metals, page 38
Unit 13: Transferring patterns, page 50
Unit 19: Cutting, page 68

Unit 14: Piercing

A small handheld saw is used to cut out, or "pierce," metal. Metal can also be cut with snips or, for larger pieces, a guillotine, although either of these will leave the cut edge quite distorted and very sharp so are not suitable for accurate cutting.

CUTTING METALS

Copper: Copper is an excellent metal to practice piercing on. It feels quite open-grained and as it is cut, the dust falls in relatively large particles. As a guideline, on 18-gauge (1-mm) sheet copper, use a 1 or 00 blade.

Brass: Brass is quite similar to copper but feels slightly coarser. When sawing nonprecious metals, catch the dust in a clean tray or skin because it should be kept separate from any precious metal dust. For 18-gauge (1-mm) sheet, use a 1 or 1/0 blade.

Aluminum: It is very important to keep saw blades for aluminum separate from those used for other metals because the dust particles would contaminate them during any heating or soldering operations. Aluminum feels much grainier and lighter than either copper or brass. For 18-gauge (1-mm) sheet, use a 1 or 2 blade.

Several grades of blades can be fitted into a saw frame. The coarsest blades are used to cut thick metal—anything over about 9-gauge (3-mm) thick—and the finest blades are used on metal less than 24-gauge (0.5-mm) thick. Nonprecious metals feel a little different from the precious metals to cut, and they generally require a slightly coarser blade to cut a similar thickness.

The cutting action should be smooth. There should be no tension in the cutting hand; hold the handle of the saw quite loosely. Hold the metal on the bench pin with your free hand. If you allow your hands and arms to become tense, the blade is more likely to break or get stuck in the metal.

Lesson 15: Fitting the saw blade

The blade is fitted correctly if it makes a "twang" like a guitar string. If it becomes loose as you are working, stop to tighten it up again.

1 Take a blade out of the packet and turn the edge with the teeth toward you. Run your index finger down the teeth. It should be smooth on the downstroke and prickly on the upward one.

2 Hold the saw in one hand and push the top edge against the middle of the bench pin. Place the top edge of the blade into the screw slot and tighten it up. Push the saw into the bench pin, and then place the bottom edge of the blade into the bottom screw slot and tighten it.

3 The blade should feel tight and springy. It is now ready to start cutting. Rub the blade with a little beeswax or a bar of soap to ease the first cut. Keep the beeswax or soap handy to keep the blade from binding.

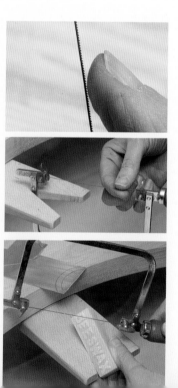

Lesson 16: Cutting a shape

When you are cutting, make sure there is only one line to follow. Double or feathery lines will result in inaccurate cutting.

1   Place the metal on the bench pin with the tracing paper design in place or with the pattern scribed on the surface.

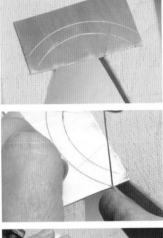

2   Hold the saw at right angles to the metal and with the blade just touching the line that is to be cut. Place the thumb or forefinger of the hand holding the metal to guide the blade to the cutting line.

3   To help the first cut, hold the saw at an angle to the metal. This helps to get it going.

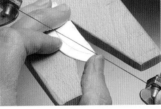

4   Straighten the saw and cut along the line with long smooth strokes. If you want to cut a curved line, keep the blade moving and slowly turn the saw or move the metal slowly around while the saw stays in the same position.

5   To cut a right angle, cut up to where you want to turn. Use the smooth back of the blade and continue the cutting motion, right into the corner. This is like treading water, keeping the motion going without moving. As the back of the blade is against the corner, turn the saw into position to cut along the next line.

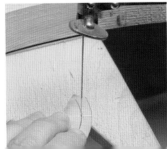

CUTTING METALS CONTINUED

Stainless steel: Stainless steel feels quite close-grained, but very hard. Sawing stainless steel takes much longer than for other metals, and the saw blades may wear out fairly quickly. For 18-gauge (1-mm) sheet, use a 1 or 2 blade.

Standard silver: Standard silver is an easy metal to cut. Pure silver is very soft and therefore even easier but will bend as you are cutting. It has quite a close grain, and the dust can be collected in the skin and saved to sell with the other scrap pieces. Use a 0 or 01 blade on 18-gauge (1-mm) sheet, a 02 or 03 blade to cut 20-gauge (0.75-mm) sheet, and anything under 24-gauge (0.5-mm) sheet should be cut with a 04 or 06 blade.

Low-carat gold: The major part of a low-carat gold is a mixture of nongold metals. This can mean that it feels slightly coarser than silver to cut, although there is nothing difficult about it. Gold dust from cutting should be kept separately, so make sure the skin is clean before you start. For 18-gauge (1-mm) sheet, try a 00 or 1 blade.

High-carat gold: High-carat gold has a close-grain feel about it and cuts beautifully. All gold dust should be stored carefully according to carat. Use a 01 or 02 blade to cut 18-gauge (1-mm) sheet, a 02 or 03 blade for 20-gauge (0.75-mm) sheet, going down to a 06 or 08 blade for 28-gauge (0.3-mm) or less.

For silver, low-carat, and high-carat gold, collect the dust or filings. Keep each separate to send to your metal dealer.

Lesson 17: Cutting a straight line for a ring

When the metal is placed on the bench pin for cutting, make sure that the light shines in such a way as to highlight the scribed line you have marked to cut along. Don't cut until you can see it clearly.

PIERCING TIPS

Freeing the blade: There will be times when the blade will simply not move backward or forward. If you try to force movement, the blade will break. Find where the blade should be by lifting the metal and the saw off the bench pin, and allowing the metal to turn to where it wants to be. Lower them both back onto the bench pin, being careful not to alter the position or angle. The blade should then run freely.

Breakage: Blades break easily. The usual reason for a breakage is that the blade is being forced to do something too hard or too quickly.

Cutting a straight line: If you find it difficult to cut in a straight line, draw a second parallel line before starting to cut, and cut between the two lines.

Cutting corners: If you are cutting out a strip for a ring, cut up the length, then bring the saw out of the work and make a fresh cut into the metal to cut across the width. This produces a cleaner corner than trying to turn the blade in the metal.

1   Make sure at least one edge of the metal is straight. If necessary, use a file to straighten it.

2   Open out a pair of dividers to the width to be cut. Place one point of the dividers on the edge of the metal with the other point on the surface. Draw the dividers down the edge of the metal, marking a line.

3   You may find it easier to cut a straight line by cutting between two lines. Draw two lines $\frac{1}{32}$ in (1 mm) apart. Cut between the two.

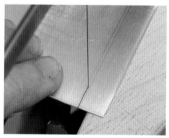

TEXTURED STRIPS
This ring was formed from straight strips of sheet metal, textured and then bent into shape.

Lesson 18: Cutting out interior shapes

You may want to cut out an area from the interior of the metal without cutting in from the edge. Drill a small hole close to the edge of the area to be cut out. If the piece is small, drill holes before cutting out the outside edge.

1   Stick the tracing paper onto the metal. (See Lesson 12, page 50.) Make a pencil mark where each drill hole should go. Put the metal on a hard surface.

2   Mark a location point for each hole using a center punch and a hammer.

3   Hold the metal down on a piece of wood. Drill small holes through the tracing paper.

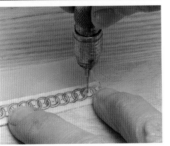

4   Undo the bottom of the saw blade, thread it through a hole, and fasten it up tightly again. (See lesson 15, page 52.)

5   Place the piece onto the bench pin and start cutting out the enclosed area. If tight corners make the cutting difficult, cut up one side, then bring the saw back and cut up the other side. It may be necessary to remove metal as you go to allow freer movement of the saw in a tricky area.

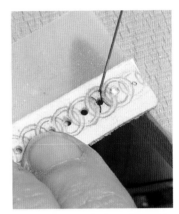

6   When you have removed the area required, gently rub the saw blade down the sides of the hole to act like a file. This is sometimes more effective than trying to use a file in very small areas.

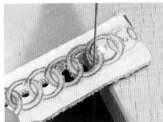

INTERIOR PIERCING TIP

It is better to do any interior piercing after an article has been bent or shaped. If a pattern is pierced into a circle of flat metal that is then shaped into a dome, some of the edges of the pattern will stick out at a different angle from the rounded shape of the dome. They will stay the correct shape if they are cut after the dome is formed. This is also true for drilled holes; they will end up slightly oval if the metal is shaped after drilling.

FEATURE JUMP RINGS
This neckpiece uses large and small jump rings very effectively to make an interesting and flexible collar around the neck.

CUTTING JUMP RINGS FROM WIRE

Cutting many jump rings: Jump rings can be made from any appropriately sized wire. If you are making more than a couple of rings, it is quickest to wind a length of wire around a former of the correct size and then cut each one away (see page 122).

Cutting tip: It helps to keep the rings in place and to cut them evenly if you turn the saw blade upside down to cut them. The pressure is then kept on the coil until the ring has been cut through.

Lesson 19: Cutting wire

Wire can be cut with snips or top cutters, but they always leave a pinched end that must be filed straight afterward. So to cut any wire over about 18-gauge (1 mm) in diameter, use a saw.

1   Mark the wire with dividers where you want the cut to be. Cut about halfway through the thickness of the wire.

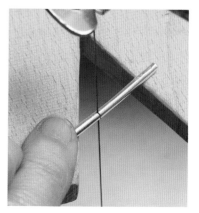

2   Push the wire away from you on either side of the cut; otherwise the metal will fold up around the saw blade, causing it to get stuck and break. Finish the cut and file any sharp edges.

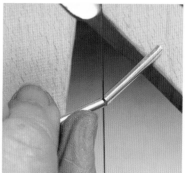

Lesson 20: Cutting thick tubing

Thicker tubing will not fit into a tube cutter (see Lesson 21) and must be cut by hand with a saw.

1 File the end of the tube flat with a large flat file. Open the dividers to the width that you want to cut off. Hold one point of the divider against the edge of the tube and use the other point to scribe a line all the way around.

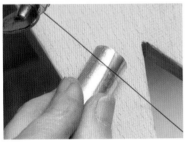

2 Place the saw on the scribed line at an angle and cut a faint line all the way around. You should finish up where you started. You can now cut either by continuing to work in the same way around the tube or by completing one area at a time until it is all cut through.

Lesson 21: Using a tube cutter

When cutting lengths of finer tubing, use a tube cutter. This little tool holds the tubing and the blade straight, allowing you to cut a neat line.

1 Make sure the end of the tubing is straight; if it is not, file it. Place it in the tube cutter with the filed end up tight against the end of it.

2 Undo the screw on the tube cutter and pull out the end until it gives the correct length of tubing to cut. Tighten the screw and place your saw blade down the slot to cut the tubing. Use the same setting to cut as many equal lengths of tubing as required.

PRACTICE PROJECTS

Use this technique to make the wire-decorated earrings; see page 132.

Unit 15: Annealing

Before a metal is worked it must be annealed—through a process of heating, it is softened and made pliable. After annealing, the metal can be worked and shaped. When it starts to workharden it must be annealed again.

ANNEALING TIPS

Deciding to anneal: When you buy a sheet of metal, always assume that it needs annealing before bending or shaping. Wire is a little different; it is easy to check whether 18-gauge (1-mm) round wire is soft, but it may be less obvious if 4-gauge (5-mm) round wire is. Always anneal thicker wire before starting work.

Pure silver and gold: Pure 999.9 silver and pure 24-carat gold do not oxidize when heated. They are intrinsically soft, and annealing only needs to take place if they have become hard through much working.

Flame size: Annealing should be done with a large "soft" flame. Precious metals are very good conductors of heat, so they will achieve their annealing temperature quicker if the flame is gently pushed up the length of the metal rather than being waved from side to side.

Protective plastic: Standard silver is usually supplied covered with a plastic coating to protect it against scratches. This coating can be kept on for some piercing applications, but make sure to remove it before any annealing.

Stay dry: After quenching, the metal should be completely dried before it is worked. Any dampness will transfer onto steel tools causing rust and pitting. Any marks made in this way will transfer onto precious metals that are subsequently used with the tool.

As they are heated, most metals change color. The first change is a darkening appearance, followed by quite an obvious blackening, known as "oxidizing." To anneal a metal, the heating process is continued through the oxidizing stage until the metal starts to show a dull red. This color is held for a few seconds as the metal relaxes; then it is "quenched" (cooled). It is almost impossible to be absolutely precise about the moment metal needs annealing. Make sure that the metal you are working with never becomes too hard. As you start working with different metals you will begin to feel the difference between their soft and hard states.

Lesson 22: Annealing thin wire

Wrap thin wire into a coil before annealing. Wire that is allowed to flop will melt more easily.

1   To anneal a coil of 18-gauge (1-mm) wire, tuck both ends back into the coil so that it will not spring apart when it is heated. Place the coil on the soldering block.

2   Heat the coil with a large soft flame. Keep the flame moving; intense heat in one area will melt the wire. Turn the coil over with a pair of insulated tweezers to anneal the other side.

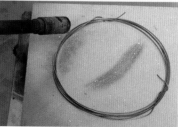

Lesson 23: Annealing thick wire

Thick wire is not as easy to coil as thin wire. Lay it flat on the soldering block and keep it supported all the way along so that it does not flop as it is being heated.

1   Place a length of 4-gauge (5-mm) wire on the soldering block. Push a large soft flame along the wire. As one area becomes dull red, move the flame up the wire until it has all been annealed.

2   Allow the wire to cool for a few seconds and then quench in water. If necessary, bend the wire so that it will fit into the pickle. Allow the wire to sit in the pickle until it is completely white. Remove the wire and rinse and dry it before working.

SEE ALSO
Unit 9: Precious metals, page 34
Unit 10: Nonprecious metals, page 38
Unit 16: Quenching and pickling, page 60

Lesson 24: Annealing a sheet of silver

To prevent "firescale" during annealing (see page 84), cover the silver with a protective non-oxidizing powder. The protective powder should not be applied over solder joins because it will make the solder run.

1   Remove the plastic coating from the silver and rub some 400-grade wet and dry paper over both sides of the silver to remove the shine.

2   Place some powder in a saucer and mix in a little methylated spirit or water to make a thick paste. Mix in a fraction more liquid so that it can be easily painted over both sides of the silver.

3   Place the silver on the block and heat it with the blue end of a large soft flame. Push the dull red color up the metal until the whole piece is annealed.

4   Allow the silver to cool for a few seconds; then quench it and place it in warm pickle for 5–10 minutes (see page 61) to remove the protective powder.

Unit 16: Quenching and pickling

"Quenching" means cooling hot metal in water. It holds the metal in a soft condition after annealing. "Pickling" is the means of cleaning the metal after it has been heated. There are several different pickles, all of which work well, although they all act faster and more effectively if they are used in a warm state.

The liquid used for quenching is cold water, held in a strong, preferably glass, container. Some metals, for example white or red golds, are not quenched immediately after annealing. They may need to cool in the air first. Your metal supplier should be able to give you technical information on quenching different metals.

After you have annealed a piece it may have a layer of oxide and a glasslike residue of flux. Placing it in pickle for a few minutes will cause the oxide and flux to disappear. This will allow you to spot errors and deal with them immediately.

TYPES OF PICKLES

Alum: Alum comes in a white crystalline form and is available from most chemists. It is officially labeled alum (potash). It is a sulfate of aluminum and was used as a medical astringent. However, for the jewelry maker's purposes it dissolves copper oxides very well.

Safety pickle: This varies according to the make; follow the instructions on the container carefully and make up enough to fill a plastic container to within a couple of inches (50 mm) of the top. The dry contents may contain some part of sulfuric acid, so handle everything with rubber gloves and in a ventilated area.

Vinegar and salt or lemon: A warm solution of salt and vinegar or lemon juice can be used to clean metal. Lemon juice works particularly well with copper. A teaspoon of salt with about ½ pint (300 ml) of white vinegar will clear any oxides from copper. These solutions may be a little slower with silver.

Sulfuric acid: Traditionally sulfuric acid was the most common pickle used in jewelry workshops. It is very effective for cleaning off oxides and flux residue quickly. However, any splashes on skin are very prickly, and splashes on clothing always result in holes. Use a solution of one part acid to ten parts water. There will be fumes given off when the acid is poured into the water, so it should always be done in a very well-ventilated area. The golden rule here is acid to water, never the other way around.

IMPORTANT SAFETY INFORMATION:

All acids are dangerous substances. If you spill acid on either yourself or a surface, make sure to rinse it with lots of cold water right away. Make a wet paste of water and baking soda and scrub the area. This will neutralize the sting of the acid on your skin. If you do not notice the spill, the following day you will have a little black mark on your hands. Unfortunately, once the black mark is there it takes a few days to disappear, and no amount of scrubbing will get rid of it.

Always use brass, plastic, or stainless steel tweezers when placing anything in or removing it from any kind of pickle.

Lesson 25: Pickling with alum

Alum works quite slowly so the metal may take a while to clean, but this pickle does not cause skin burning or holes in your clothes when spilled, making it the safest one to use.

1   Place 2 or 3 tablespoons of alum crystals into a plastic container. Add enough warm water to come within 2 in (50 mm) of the top and stir.

2   Place the plastic container in the water inside a slow cooker and switch the cooker on. Allow it to become quite warm; then turn the temperature down to low. The water outside the container should reach about halfway up the side.

3   Anneal a piece of silver or copper (see page 59), quench it, and then place it in the warm alum solution.

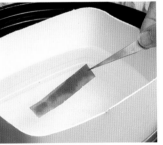

4   Leave the metal in the alum for at least 5 minutes or until it is completely clean. With silver, this will leave the metal looking very white. With copper all the firescale will disappear, leaving a light shiny pink surface.

5   The alum will be active for a few weeks, but with time it does deteriorate. It will become a dark blue color and the metal it is supposed to be cleaning will stay rather gray. The spent alum can be flushed away, with lots of running cold water.

PICKLE HEATING TIP

To keep your pickle hot, use a slow cooker. (If you buy a used cooker, get the wiring checked before you use it.) Half-fill the cooker with water. Cut the bottom half from a plastic milk carton and fill it with pickle to within 2 in (50 mm) of the top. Place the plastic container of pickle into the water and plug the cooker in. Keep it on a low heat all day so that your pickle is always warm.

SEE ALSO
Unit 15: Annealing, page 58
Unit 21: Soldering, page 72
Unit 24: Polishing and finishing, page 84

Lesson 26: Pickling with sulfuric acid

Only use sulfuric acid for pickling if there is no alternative. You may need to mix solutions of other acids, in different proportions, for different uses; the same safety steps apply. Above all, do everything slowly and carefully.

1 You will need a pair of rubber gloves, a face mask, a glass measuring jug, a plastic bowl, cold running water, sulfuric acid, and an open window or extractor fan. Fill the measuring jug with ten parts of water and place it into the plastic bowl.

3 Gently stir the solution with a glass or plastic rod and carefully pour it into the plastic container to go in the slow cooker.

2 Put on the rubber gloves and undo the lid of the sulfuric acid, either under the open window or with the fan on. Gently pour the acid down the inside edge of the measuring jug, until you have added one part acid to ten parts water. Put the lid back on the acid and store it safely, preferably in a locked cupboard.

ACID SAFETY TIP

Always wear safety goggles when using acid. Be very careful when taking the lid off the slow cooker if your pickle is acid. Do not lean over it right away because the fumes are very strong. Do not put hot metal directly into the acid. It will make it spit, resulting in hot acid droplets landing, if not directly on you, on surrounding surfaces and tools.

Unit 17: Cleaning

Keeping a piece clean as it is being worked is enormously helpful in keeping mistakes to a minimum, keeping measurements accurate, and knowing at each stage of construction what is happening and what needs more attention. Keeping things clean not only applies to the metal itself, but also to work surfaces, tools, and soldering equipment.

CLEANING TIPS

Protective plastic: Leave protective plastic on metal for as long as possible. Remove scratches with a file or wet and dry paper before they become too difficult to reach during construction.

Pickle: After each soldering and annealing, pickle your piece. Leave it long enough for the oxides and any excess flux to disappear.

Pumice: Keep a saucer of pumice powder and an old toothbrush by your sink. Wet the toothbrush with water, dip it in the pumice powder, and rub this over your work. Don't use it on a finely finished surface.

Brass brush: A soft brass brush is a useful item to keep by the sink. Pour a little liquid soap onto the brush and rub vigorously to create a bright shine. It will cause scratching on a finely finished surface.

Liquid silver cleaner: To clean a tarnished silver chain, place it for a few seconds in a liquid silver cleaner. Remove it as soon as it is bright because overexposure in these liquids can dull items. Wash in liquid soap and water to remove any stickiness. There are similar liquids for gold.

Impregnated cloths: There are commercially available cloths that are impregnated with a cleaner. Fasten a corner of the cloth in the vise on your bench and pull it tight. Rub the article to be cleaned on the cloth. It will bring it up to a lovely shine.

Acetone: Polish can also be removed with a little acetone soaked into a soft cloth. A good alternative to acetone is nail polish remover. Polish can also be removed with a weak solution of ammonia.

Bear in mind that your work may oxidize after it is finished. Sometimes, this can be used to good effect. Copper, for example, may take on a green or dark brown color. Silver, unless it has been treated, will become rather tarnished, but it is fairly simple to clean.

Lesson 27: Cleaning off polish

The cleaning method shown here is a cheaper alternative to using an ultrasonic cleaner.

1   Use a gas or electric ring and an old enameled saucepan. In the saucepan, mix a squeeze of liquid soap, a dash of household ammonia, and 2 cups of water. Place the pan on the ring, boil the mixture, and then lower it to simmer. Place your item in the mixture.

2   As the liquid bubbles, the item will move around in the bottom of the saucepan. This vibration helps remove the polish, so let it simmer for a good 10 minutes. Turn off the heat and remove the item from the pan with a pair of stainless steel tweezers. Rinse it under the tap.

Unit 18: Bending

Metal must be in an annealed state to bend and stretch easily and uniformly. Some metals stay relatively soft as they are being worked, but eventually, all metals harden while being hammered, or worked over metal stakes, or put through the rolling mill.

CURVED BROOCH
A strong but simple curve of bent metal defines the shape of this restrained brooch design.

DECORATED RING
The decoration for this chunky ring was made from a wide silver strip, textured, bent into shape, and then soldered onto the ring shank.

STUD EARRINGS
These gold earrings were formed from four domed disks which were cut to the center, curled, interlocked, and soldered.

CURLED EARRINGS
The drop sections of these gold earrings were formed by curling textured strips around a former.

Some bending can be achieved by simply manipulating the metal with your fingers to create the shape you want. If the shape requires more intricate bending, then stakes, formers, mallets, and hammers are used.

Because the metal is in a soft state for shaping, it is also vulnerable to unwanted marks made by the tools used to bend it. A huge range of tools is available for creating different shapes; however, their correct use is essential. Steel tools mark softer metal if they are used incorrectly; wooden tools only mark very soft metal such as fine silver or very high-carat golds.

SET OF RINGS
This set of rings shows very good use of neat sharp bending. The sharp bends were made by filing a groove on the inside of the bend first so that the metal closed at the correct angle when bent with the pliers.

Lesson 28: Bending sheet metal into a ring

You will need a strip of annealed metal and a pair of half-round/flat pliers for this lesson. This method is used to form a ring.

1   Use a pair of half-round/flat pliers and hold the end of the strip of annealed metal between them. The curved side of the pliers will form the inside of the curve of the ring. Push the metal around the pliers until the metal has formed a U-bend.

2   Turn the metal around so that you can hold the other end of it in the pliers and repeat the action in reverse, bringing the second end round in a U-shape to meet the first.

3   Line up the two ends so that they sit tightly together by pushing them past each other and then bringing them back together. Solder the join (see page 74). Examine both sides of the joint. You should see the silver solder seam easily.

4   After soldering, put the ring onto a mandrel, and hammer it with a wooden mallet until it is round.

Lesson 29: Bending wire into a ring

You will need a length of small round wire and a pair of round/flat pliers for this lesson. This method can be used to make a single jump ring.

1   Hold one end of the wire in a pair of round/flat pliers. The inside diameter of the small ring will be a little bigger than the outside diameter of the round bit of pliers. Wrap the other end of the wire around the pliers to form more than a complete circle.

2   Remove the wire from the pliers. Hold the loop of wire on the bench pin and cut through both thicknesses with a saw.

3   Close the circle with two pairs of flat-nose pliers in a sideways movement.

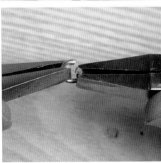

Lesson 30: Making angles

To make a right angle or a more acute bend in annealed sheet or wire, you will need a pair of flat-nose pliers and a triangular or square file.

Lesson 31: Making spirals

Use this method for wire of 18-gauge (1-mm) diameter or less.

1   Mark with a pen or scribe where the bend will be. Use the edge of the file to make a groove in the metal until it is about halfway through the metal thickness.

2   Hold the metal, close to the groove, with the flat-nose pliers. Use the thumb of the free hand to gently but firmly push the other side of the metal over to a right angle. If you want more than a right angle, file the groove a little deeper. When the angle is correct, and the sides touch tightly, solder the groove closed.

1   You will need a pair of round-nose or round/flat pliers and a length of annealed wire. Turn the pliers with a circular movement while holding the wire tightly to push it around into the first small circle.

2   Remove the pliers and replace them further along the length of wire while you continue to curve. Keep the wire just at the tip of the pliers so that the spiral keeps the same width. Continue until the spiral is the desired size. Trim the first curve with the saw.

USING BENDING TOOLS

Leather rawhide, wooden, or plastic mallets: Mallets are used to shape metal without marking it. For example, a soldered ring that needs to be made round is placed on a steel mandrel and hit with a mallet until it is the correct shape. The outside of the ring will not be damaged by the mallet striking it.

Round wooden dapping punches: If you have a patterned or textured metal you want to make into a dome shape, use a round wooden punch in the metal dapping block. The wood will not distort the texture or pattern.

Steel hammers: Steel hammers stretch and shape metal, but they leave marks. A steel hammer with a clean and polished flat edge only leaves light marks, but a rounded hammer, or one with any marks on the head, transfers those marks onto the softer metal.

Steel stakes: Stakes are used to support and shape the metal as it is being worked. Keep them clean and polished so they will not mark the metal on the underside. Make sure the hammer that is working the metal from on top never strikes the steel stake; it should only strike the metal being worked.

Safe-jaw vise: A safe-jaw vise is usually fixed to your bench and can be used to hold steel formers without marking them. The safe jaws are made from plastic or rubber. If you only have access to a vise with serrated metal jaws, you can make a pair of safe ones with either aluminum or copper bent at right angles to fit over each edge.

INTEGRAL CATCH
The fastening in this brooch has been cleverly incorporated into the design of the whole piece. The catch, made by careful twisting with round/flat pliers, is an integral part of it.

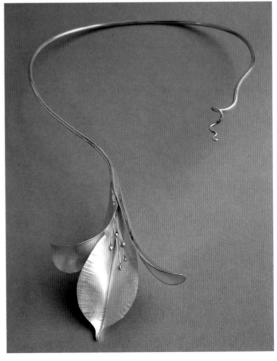

TENDRIL DETAIL
Although this neckpiece does not need a fastening, the end has been given a tendril-like appearance with just a few turns of the pliers.

SEE ALSO
Unit 15: Annealing, page 58
Unit 21: Soldering, page 72
Unit 23: Using stakes, page 82

Unit 19: Cutting

There are ways of cutting metal other than with a saw. Use a saw if you are cutting anything from sheet metal because it leaves the neatest edge and is really just as quick as trying to flatten and smooth out an edge left by the guillotine. However, consider using the guillotine if you have a really large piece of sheet, especially if it is more than 18-gauge (1 mm) thick. Sheet metal can also be cut with shears with one handle held in a floor standing vise, so that only the other handle needs to be pushed down against the metal to cut up a long strip.

TOP AND SIDE CUTTERS

Top and side cutters can be used to trim ear wires or any pieces of wire that have been soldered through a hole. They are designed to get close enough to just leave a little top, which can then be filed. Top cutters can also be used to make a small indented line around the wire post on an ear stud (right). This line helps the butterfly backing to stay on.

Other commonly used cutting tools include snips, top cutters, side cutters, scalpel, craft knife, and scissors. Scissors, of course, are used for all paper and material cutting; the scalpel or craft knife is useful in cutting card, silver and gold foil, rubber, and also for lifting away excess epoxy resins if necessary. Snips can be used to cut along thin metal, such as a copper or tin scrim and thin silvers and gold. They are also used to cut binding wire, solders, and other thin wires. They will not cut clean anything much thicker than about 24-gauge (0.5mm).

CUTTING WIRE
Wire cut with top cutters always has a squeezed end, which must be filed straight before it can be used.

SEE ALSO
Unit 8: Essential tools, page 26
Unit 21: Soldering, page 72

Lesson 32: Cutting silver solder paillons

Silver solder is used to solder copper and brass as well as silver. Silver solders come in thin strips or wires.

1. Thin the strips before cutting paillons. Either put them through a rolling mill or hammer them. Leave one end untouched, and scratch an identification mark on this end: EN for enameling solder, H for hard, M for medium, E for easy, and XE for extra easy.

2. Use snips to make several cuts along part of the length of the thinned solder.

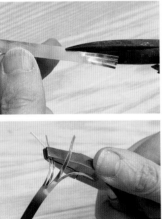

3. Because these cuts will start to curl up, use a pair of flat-nose pliers to straighten them out.

4. Hold the solder and use the top cutters to cut across the strips. Hold your hand underneath as you cut so you can catch all the little paillons.

Lesson 33: Cutting gold solder paillons

Gold solder comes in pennyweights (dwt) and is thinner than silver solder. It doesn't need thinning unless very tiny paillons are required. Gold solder is marked by carat by the manufacturer.

1. Use snips to cut up a small area of the solder.

2. Use the top cutters across the strips to cut very tiny paillons. (Gold solder flows better if the paillons are placed little and often.) Store the paillons in a plastic container with the appropriate label.

🔊 SOLDER PAILLONS

A "paillon" is a thin leaf or chip of metal, and in this context refers to very small pieces cut from any solder. Find some little plastic containers into which you can put paillons of different solders so that they will always be on hand. Stick a small piece of masking tape onto the lid of each container and mark it with the name of the solder inside to identify it.

Unit 20: Joining

There are basically two ways of joining one piece of metal to another. One way is to make a mechanical joint, which allows for movement in one or two directions. This could be a rivet, screw, hook, hinge, jump ring, or any other variation on this theme. The other, permanent, way of joining metals is soldering, covered in Unit 21 on page 72.

Necklaces, chains, bracelets, and pendants often have movable joints in at least one part of them. A joint is an integral part of a piece and should be part of the overall design. Careful consideration must be given to the sort of movement required in the piece being made. For example, in order for a necklace to sit comfortably around the neck, the links should be joined in such a way as to allow movement in two directions.

RIVETED NECKPIECE
This simple piece drapes around the wearer's neck.

Lesson 34: Using jump rings

A popular way of joining links is with a "jump ring." A jump ring is a circle made from wire, which can be slotted into two holes to join one piece to another.

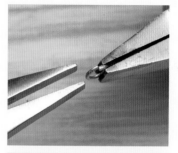

1   To open the jump ring, take two pairs of flat-nose pliers and hold a pair in each hand. Hold one half of the jump ring in the first pair, and hold the ring just the other side of the join with the other pair.

2   Twist one pair slightly away from you and bring the other pair slightly toward you. This will open the jump ring without losing the shape of the circle.

TYPES OF JOINING LINKS

RIVETS
A rivet is used to hold pieces together without solder. Push a small piece of wire through similar holes in two or three separate pieces. File the top and bottom faces flat and hammer them to spread them out (see page 102).

SCREWS
There is a special "tap and die" set made for jewelers, so very small wires and tubes can be used to make screws that are in proportion to the piece. The size of the wire to be tapped should be larger than the inside diameter of the tube but smaller than the outside.

3 Keep hold of the jump ring with one pair of pliers and slot two links onto it. Now use both pairs to reverse the opening movement and close the jump ring. Solder the join (for more on soldering, see page 72).

SEE ALSO
Unit 21: Soldering, page 72
Unit 27: Fittings, page 92

JOINING TIPS

Mock up links: When making joining links for a new design, it's a good idea to make up a couple first as a mock-up (see page 22). It will tell you all you need to know about how well the links move, whether they are too tight or do not hang well, or if the look is what you want.

Play with ideas: Joining links by having holes and jump rings can be a little predictable. Try to find a different and maybe more interesting way of using the principle.

Make links strong enough: Make sure that the wire used for a jump ring is strong enough for the piece. If it is too light for the weight of the chain or necklace, the circle will distort.

Make ovals: To make a round jump ring oval, solder the join, then hold the tips of a pair of round-nose pliers inside the ring. Gently open out the pliers so the circle is pulled out into an oval.

Solder jump rings: Jump rings always look neater if they are soldered. If you make sure the join is really tight, you will only need a tiny piece of solder to hold it together. If the solder is too big, it will show as a lump and will have to be filed down.

HOOKS
Use a hook to fasten a chain or a necklace, or to hang extra pieces from, providing there is no chance it will unhook itself during wear. For example, you may want to hang a small pendant from a simple chain, but keep the option of having the chain on its own.

HINGES
Use a hinge to join two pieces, which can then be moved without losing the integral line. Interlocking pieces of tubing are soldered alternately onto the edge of each piece and are joined with wire going down through all the tube sections.

Unit 21: Soldering

A permanent join between pieces of metal is achieved by soldering. Solder is an alloy of the metal that is being joined, which includes a metal such as zinc to help lower the melting point. A hard solder contains less alloy than an easy solder, so hard solder will not flow when easy solder is applied on the same piece. The different solders and their melting points are covered in Unit 9, page 34.

The solder must flow below the melting temperature of the metal, so this affects which solder is used with different metals. Solder needs a medium to make it flow—the "flux." Flux comes in liquid, powder, and cone forms. The dry forms are mixed with a little water to make a paste; the liquid is applied directly.

Set up your soldering area in a dark corner of the workshop (see page 24). Keep a light handy so you can see where to place the flux and solder, but turn it out as the torch is lit. As the metal heats up, it changes color, and as the solder is about to flow the metal glows brightly. If there is too much light it is impossible to see this.

Try to solder as much as possible with hard solder before moving on to either medium or easy solder. If you use hard solder for subsequent joins (after hard solder was used on the first), paint the previous joins with some flux because this allows the solder to flow again if it wants to, rather than burning out, which can happen with repeated heating at high temperatures. When solder burns out, the join looks lumpy and the solder line stays oxidized for longer than might be expected after pickling.

CHAIN-LINK BRACELET
This bracelet incorporates a variety of cabochon stones to enhance the handmade chains, soldered to end pieces that form the closure.

SOLDERED DETAIL
The gold decoration on the top of this striking brooch was soldered onto its curved silver frame. The central panel would have been placed after soldering.

PRACTICE PROJECTS
Use this technique to make the wire-decorated earrings; see page 132.

SEE ALSO
Unit 7: The workbench, page 24
Unit 9: Precious metals, page 36
Unit 18: Bending, page 66

Lesson 35: Testing different solders

You may have unidentified strips of solder on your workbench.
Do this simple test to find out which is which.

1 Find a little piece of scrap
silver and paint on a dot of
flux for each strip of solder
to be identified (in this
case, three).

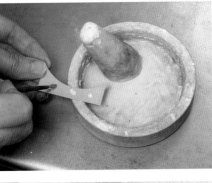

2 Lay the pieces of solder
you are unsure of in a line
so that you can identify
them later. Cut a paillon from
each one (see page 68) and
place these on the flux in the
same order.

3 Now introduce the heat
slowly to let the flux settle
down and then watch to
see the order in which each
paillon runs. This happens
rather quickly so watch
carefully! The one that melts
first is the lowest-melting
temperature solder (easy),
and the hard solder will be
the one that melts last.

SOLDERING TIPS

Keep your work clean: Anything that is dirty or is
oxidized will not solder.

Close up gaps: There should not be any gap between
the pieces. Solder flows well in tight spaces.

Always use flux: The join must be fluxed for the
solder to flow correctly.

Place paillons carefully: The paillons of solder must
be placed directly onto a simple join (from behind or
the front) or under the join as it sits on the soldering
block. When soldering one piece to another, the
paillons must touch both.

Apply enough heat: Very often, the main reason
for solder not flowing is that the item is not hot
enough. Providing the join is close fitting, flux has
been applied, and the whole piece is hot enough, the
solder will run.

Soldering onto a texture: When soldering an item
such as a bezel onto a textured surface, place the
solder on the inside of the item. This prevents the
solder flowing out and spoiling the texture.

Soldering a large piece: To solder silver, the whole
piece must be heated to soldering temperature. For a
large piece, such as a bangle, make an "oven," using
soldering blocks or firebricks, for the back part. The
heat can then be distributed evenly on either side of
the join while the rest of the bangle will retain the
necessary heat.

Lesson 36: Soldering a ring join

Use this technique to make a neat join in a ring. Keep the solder on the underneath or the outside of the join. For a ring with a textured surface, put the solder on the inside.

SOLDERING TIPS CONTINUED

Soldering jump rings: To solder jump rings used to join the ends of a necklace, place two small pieces of mica on either side of the jump ring and over the finished necklace pieces, and use a small but hot flame to solder the jump ring.

Pickling: Pickle your piece really well after each soldering operation. Have a look at it through a x10 loupe to check the join is a good one. Do any resoldering if necessary.

Starting again: Always check to make sure that the piece being soldered will not move as it is being heated or touched with tweezers. If a solder looks to be going wrong, stop it, quench and pickle it, and start again.

Soldering flat pieces: To solder flat surfaces together, place surface A upside down on the soldering block and cover it with flux. Place several solder paillons over the surface and heat it up until they run. Quench and pickle. Remove from the pickle, and rinse and dry and remove the top of the "bumps" of solder with a flat file. Paint the surface with flux again and put it on top of surface B. Heat up the pieces, aiming the flame more toward B, and watch for the shiny line of the solder to appear around the edges of A. You may have to push down slightly on A, to remove trapped air or too much flux.

Annealing: To solder a join after a piece has been hardened by hammering or bending, anneal it first. If it is not annealed, the introduction of the heat for soldering will cause the join to open up as the metal becomes relaxed. It will then be impossible to solder the join.

1 Cut a strip of metal and use a pair of half-round pliers to bring the ends of the strip together (see page 65). The ends should be parallel, which will make sure that they sit closely together. Dip the borax cone into some water and grind it around in the ceramic base to make a paste. Use a paintbrush to apply a little paste on, and through, the join.

2 Dip the paintbrush into the borax again and lift up a paillon of hard solder on the tip of the brush. Place the solder paillon onto the soldering block.

3 With the join of the ring facing toward you, place the center of the join on the solder.

4 Gently bring in the flame and heat the ring all around, allowing the flux to bubble and then settle. Increase the heat until the metal is a strong red color and the solder flows up the join—it will become a bright shiny line. Allow the line to develop along the join, before removing the flame.

PRACTICE PROJECTS

Use this technique to make the cabochon-set ring; see page 126.

Lesson 37: Soldering small fittings to larger pieces

Fittings such as brooch-pin hinges and catches, ear pins, cuff link hinges, and so on, are usually soldered on last, after everything else is in place, with easy solder. Some fittings will balance in their soldering position, and others such as ear pins are placed as the solder runs.

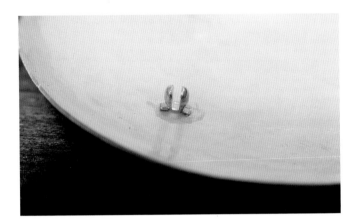

2   When soldering a fitting that is difficult to balance, such as an ear pin, paint a tiny bit of flux and place a paillon of easy solder where the pin is to be soldered. Paint some flux on the bottom of the pin and hold it with a pair of insulated tweezers. Support your elbow on the soldering table and lower the pin over the work using your elbow as a hinge.

1   For a brooch-pin fitting, place the article upside down on the soldering block. Paint a tiny bit of flux where the fitting is to go, place the fitting on top of the flux, and put a small paillon of easy solder either side of it. The paillon should be touching both the fitting and the base to which it is being soldered. Keep the flame away from the fitting by gently heating one side of the piece itself. The heat will gradually transfer to the side with the fitting. When you see the solder starting to run, gently bring the flame over to that side until the solder is complete.

3   Heat the piece up gently until the solder flows. Bring the pin down so it is right above the join and lower it into the molten solder. Hold it steady until you see the solder flow around the base, and then remove the flame while holding the pin in place. Allow to cool for a few seconds, and then pickle.

PRACTICE PROJECTS
Use this technique to make the cast cuff links; see page 129.

Unit 22: Filing

A file is used to remove excess metal or solder from a piece that is being worked on. Files are quite specific; using the right files in the right order makes a job much easier. Making the best use of a file also depends on the way it is held and moved. Support the piece you are filing on either the bench pin or the bench itself because the resistance that this gives increases the effectiveness of the file.

Files are usually supplied without a wooden handle. These can be purchased separately if you wish and fitted onto the tapered end of the file.

To fit a handle, hold the file in the safe jaws of the vise with the end protruding. Place the wooden handle on the tip and use a wooden mallet to push it down onto the file.

FILED FINISH
A fine file, followed by wet and dry papers, and finally pumice powder, were used in the finishing of this smooth stylish brooch, made from silver, 18-carat gold, and a single diamond.

PRACTICE PROJECTS
Use this technique to make the circular brooch; see page 136.

SEE ALSO
Unit 24: Polishing and finishing, page 84

Lesson 38: Filing a straight edge

A file really only cuts in one direction. When filing a straight line it is better to use only the forward stroke as the cutting stroke. Keep a flat file parallel to the work, being careful not to drop it at either end because this results in rounded or lower corners.

FILING ON THE BENCH PIN

1   This piece of silver has had previous cuttings taken from it, so the edges are not straight. To cut a strip for a ring shank, the edge needs to be nice and crisp, so that the dividers can be used to draw a line parallel to it.

2   Place the silver on the bench pin and use a large flat file on the edge so that it runs parallel to the side of the bench pin.

3   Continue filing until you think the edge is straight. Hold it up to the light with the file against it to check for gaps. Once the edge is straight it can be used to mark the line for the ring shank. (See page 54.)

FILING IN THE VISE

1   Use a square to scribe the line you want to file down to. Place the silver in the safe jaws of a bench vise and check that the scribed line is parallel to the line of the vise.

2   Hold a large flat file in both hands and use a straight forward stroke to start filing the edge. Be really careful not to let the file drop at the start or end of the movement.

FILE CLEANING TIP

After constant use, files can become clogged. They can then be cleaned out with a "file cleaner." Special files for wax are available, and because wax can be quite tricky to remove from a fine file it is worth having one of these if you intend to work with wax.

FILING TIPS

Order of filing: Use files in the right order. They go from coarse to fine. Each time a finer file is used, it removes the scratches made by the previous one.

Time: Filing can take time. First make sure that the area you want to file could not be removed by using the saw, which is much quicker.

Difficult corners: A fiberglass stick can be filed to fit into a particular corner or area that is difficult to get to. It is used to file difficult areas, such as around settings and between wires, and is just rubbed along the surface.

Using a straight edge: It is all too easy to dip the file at the start or end of the motion. To avoid this when filing a straight line, hold the work so the edge of the bench pin or the bench is parallel to the line you are filing.

Using the vise: Work to be filed in a straight line can be held in the safe jaws of the vise. Hold the flat file in both hands and move it forward, keeping it parallel to the top of the vise. Lift it up before filing forward again.

Keeping filings: Collect copper, silver, and gold filings separately. The gold ones are the most valuable, so make sure to keep these carefully. Send the silver and gold filings to a refinery.

Lesson 39: Filing inside a curve

Use either a half-round or oval file to file the curve on the inside of a ring, bezel, or bracelet. The size of file depends on the size of the piece; use quite a small file for a ring and a larger one for a bracelet. But if, for example, a lot of excess solder is on the inside of a ring, use a large file to remove most of it before moving down to a smaller file.

1   If the inside of the ring is clean, the part that needs filing will be the solder join. Hold the ring on the bench pin so the file can pass through it.

2   Place the file to the left of the solder seam with the right-hand side of it tipped downward.

3   Pass the file through the ring over the seam, gradually turning it so the stroke is finished on the right-hand side of the seam, with the left-hand side of the file tipping down. Repeat this action. Turn the ring around and file the other side in the same way.

USING DIFFERENT-SHAPED FILES

Different-shaped files are appropriate for different applications, sizes of pieces, and level of finish you require. The file you choose will also depend on whether you are filing the inside or the outside of a curved piece, for example.

Pierced curves: For a pierced-out piece with an edge that curves in and out, you will need both a small flat file and an oval or half-round one. The edge will have marks left by the saw, which can be removed by filing.

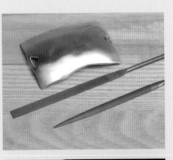

Inside curves: On the curve that goes into the work, use an oval or half-round file, with the same action as for the inside curve of a ring (see opposite).

Outside curves: For the outer curve, you use the small flat file and with a straight-over action until it cannot reach the next inner curve.

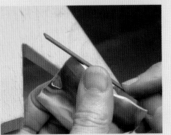

Circle edges: To file the edge of a pierced-out circle, always use a flat file. You can work down to a scribed line in this way.

Angled edges: To file an angled edge, hold a flat file at an angle of 45° against the edge of the piece and use a smooth forward stroke to make a neat edge.

Inside bends: Use a triangular file to get into small areas that need filing and to help create a bend in wire or sheet metal. File a groove just over halfway through the metal, and then bend it carefully up, so it forms a right angle.

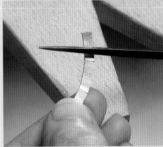

Right angles: Use a square file to file a right angle. You can also use it in the same way as the triangular file to make the groove for bending a right angle.

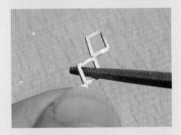

Grooves and holes: Use a round file to file a curved groove for tubing or to help open out a round hole. Be careful not to file an unintentional groove when filing in a hole.

Unit 23: Using stakes

A stake is a solid wooden, plastic, or metal shape used to stretch and form metal into three-dimensional shapes. Annealed metal can be bent over or around a stake by hand or with a mallet or hammer. Stakes range from those used to help form large items such as teapots, bowls, and tableware to much smaller items of jewelry such as rings, bangles, and necklaces.

For a stake to be effective, a hammer is used to stretch and bend the metal over it. These tools should be kept as clean as possible. Remember that any mark made by a hammer on the surface of a stake or anvil will leave a corresponding mark on softer metal when it is worked over it. Sometimes, a big old, rusty anvil is just what is needed when giving a soft metal an interesting texture, but on the whole it is better to keep the surface as smooth as possible.

The head of the hammer or mallet being used should have a firm and direct contact on the metal. The metal in its turn should be held flat against the stake so there is a direct line of contact between hammer, metal, and stake.

ANVILS

An anvil has a flat metal surface and a cone-shaped protrusion. Anvils come in many different sizes. It is useful to have a large, fairly heavyweight one, fixed to a floor-standing wood log as well as a very small one that will stay on the bench and is useful for all kinds of small flattening and shaping operations.

PENDANT AND EARRINGS
This silver pendant and earrings were given their three-dimensional curving forms by being hammered over stakes.

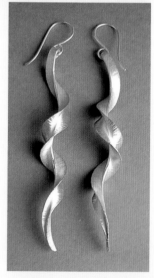

USING FORMERS AND STAKES

Holding the stake: Make sure the stake is held firmly by its tang in either a stakeholder (which is fixed to the bench), or in the vise, or that the tang is wedged into a hole in either the bench or a floor-standing wooden block. If the stake does not have a tang, wrap some masking tape or a large cloth around it and hold it in the safe jaws of the vise.

Caring for stakes: Keep stakes shiny, clean, and dry. Always dry metal thoroughly before placing it on a stake, because any dampness will transfer onto the stake and if it is not dried right away, it will leave a rusty mark. If a stake does become dull and marked, wet and dry paper can be used to remove marks and it can then be polished with white diamond polish. When not in use, a thin coat of oil can be wiped over the metal surface, to help keep it free from moisture.

Forming a ring: Another name for a round tapered "mandrel" is a ring stake. To make a ring round after it has been soldered, it is dried and then pushed down the stake by hand as far as it will go. A wooden or rawhide leather mallet, which will not mark the outside of the ring, is then used to form the metal to the shape of the mandrel.

Forming a curve: Metal will follow the curve of a stake if it forms a direct contact with it. For example, if you have a flat length of metal and hold the end on the cone part of the anvil and gently hammer it, it will start to form that curve. The curve can then be continued by feeding the length of metal around the cone and hammering it where it is touching.

Using a former: To make many jump rings of the same size, wrap wire around a metal "former." This is a rod of an appropriate diameter, one end of which is held in the vise so the wire can be formed around the protruding length.

SEE ALSO
Unit 25: Dapping and swaging, page 88
Unit 29: Hammering, page 101

Lesson 40: Making a thick ring

You may want to make a ring using thick round wire. Wire with any diameter greater than 8-gauge (3 mm) will be very difficult to bend into a circular shape with pliers, and they may leave deep marks in the metal. However, if you use a former to help bend the metal, you will find you can make rings from thick wires quite comfortably.

1   Cut the length of wire required and anneal it. Place a round former horizontally in the safe jaws of the vise, in line with the top edge. Hold the wire vertically and fasten one end of it between the former and the vise.

2   Fasten the vise up tight, and with your hands behind the wire, push the wire over the former until it is touching the front of the vise. Use a wooden mallet to push it right down.

3   Remove the wire and fasten the other end between the vise and the former. Bend the wire as far as it will go over the former. Remove the former from the vise, and push the two curved ends of the wire together by tightening the vise around the ring until the ends meet.

4   When you remove the ring from the vise, the two ends will spring apart. Because the ends must be close together for soldering, you will need to hold one half of the ring in the side of the vise.

5   With a pair of parallel pliers, hold the other half and bring it toward you and then push it toward the vise and past the other end.

6   Now bring it back and push slightly away from you to close the two ends together.

Lesson 41: Making a dome

A dome can be formed in several different ways. Smaller domes are made using the dapping block with metal or wooden dapping punches (see Unit 25, page 88). Large domes can be sunk into a floor-standing wooden block. Domes can also be hammered into shape over a metal stake, as shown here.

1   Take a circle of metal approximately 0.7 mm thick with a 50-mm diameter. Anneal and pickle it (see page 58). Hold a round curved stake in the vise, and place the outside edge of the metal circle on the sloping side edge of the stake.

2   With the flat face of a clean planishing or forming hammer making a direct line of contact on top of the metal and stake, turn the edge slowly around the stake while gently tapping the metal as it moves around.

3   When the piece has done a full circle, lower it so the area just below can now be worked in the same way. Continue until the dome shape appears. The metal will now need annealing. It can then be placed over the stake to assess the shape. Do any more hammering on the slope of the stake in the same way.

DOMED CIRCLE THEME
This necklace uses domed circles as its main theme. The smaller domes have been hammered with a ball-peen hammer, and the larger ones have a contrasting wave of high-carat gold.

HAMMERING TIP

A metal hammer over the mushroom stake (as shown here) will workharden the metal quickly. Use a leather rawhide or wooden mallet or a dead-blow hammer if you wish to form the metal more deeply.

Unit 24: Polishing and finishing

There is always demand for different and interesting finishes on metal jewelry. Silver and gold can look absolutely wonderful when given a satin or matte finish, although to achieve this look takes a similar amount of time as giving a piece a high polish. Very high-quality polishing and finishing, for a more traditional look, is also consistently popular.

FIRESCALE TIPS

Polish: Firescale can be removed from a fairly thick piece by polishing. Tripoli or general green polish will cut through the gray layer to leave a completely clear surface.

Filing or grinding: The gray firescale found on sterling silver can be removed by filing or very carefully grinding the surface.

Water of Ayr stone: This is a very fine solid stone that comes from Ayr in Scotland. Dip the stone in water and rub it into a paste over the firescale. Wipe the paste away to see if the firescale has disappeared. Continue rubbing until it has completely gone.

Bringing the fine silver to the surface: If the silver has a textured finish, there is probably no need to polish it. Bringing the fine silver to the surface is another way of removing the firescale. Heat the silver up to the annealing point (see page 59). Quench and pickle at least three times. Now brush the silver lightly with a new soft wet brass brush or very fine steel wool and liquid soap. Rinse and dry the piece. Repeat this process three or four times to build up the fine silver layer on the surface.

Protection: Mix protective antioxide powder into a paste and paint it onto silver before the first annealing or soldering. Don't paint the paste too close to the solder line, and leave a break between the flux and the antioxide paste. This is to keep the solder within the fluxed area. Reapply with every annealing or soldering.

Plating: When a piece is finished, if there are still firescaled areas remaining, the whole thing can be silverplated. This is usually done by a specialist plater. The ingredients for plating contain cyanides, so they should be treated with the utmost caution.

Much of the process of polishing and finishing a piece is concerned with restoring a fine surface to the metal and, in many cases, eradicating the appearance of firescale. The essential thing to remember is that as you remove one scratch, you replace it with another, finer one. The idea is to work down to the very finest so that the surface is almost mirrorlike.

Firescale is the name given to the oxidization of silver, caused by the copper content within it. When copper is heated it oxidizes, or goes black and forms a light black scale. Consequently, when standard silver (75 parts copper to 925 parts silver) is heated, the copper content oxidizes and turns the metal black. If the silver is then pickled it loses the black and turns quite white (see page 61). In subsequent heatings and picklings, the silver turns less and less black, until it does not oxidize at all. This is because copper is removed from the surface each time. If the piece is overheated though, the fine silver layer built up with each pickling will be burned through and oxidization will reoccur.

The fine silver layer can also be broken by filing and polishing. This means that after giving your piece a final polish, you may notice some rather dull gray patches, or your piece may be completely covered with firescale, only being bright and shiny at the edges.

POLISHING WITH THE FLEXSHAFT MOTOR
You can fit small-diameter tools to the flexshaft motor to carefully polish fine details.

POLISHING AND FINISHING TOOLS

To achieve a good finish—whether matte or shiny—all scratches should have been removed from the work. This can be done with small hand tools or with abrasive tools in the flexshaft motor. They should be applied in descending order of abrasiveness, so that each new tool or paper removes any scratches made by the previous one. Be careful using the abrasive tools in the flexshaft motor to clean a flat surface. Because of the small diameter of the tool it can very quickly make grooves in the surface rather than keeping it nice and flat.

Emery sticks: Make some "emery sticks" with different grades of wet and dry paper wrapped around ruler-length pieces of flat wood. Use double-sided tape to stick the wet and dry papers to the wood. They will be invaluable when preparing flat surfaces for polishing.

Steel burnisher: Use a steel burnisher to polish the edges of bezels and to give a fantastic shine to any finished edge. Apply it by rubbing it along the metal and pressing firmly down on it.

Polishing threads: Polishing threads are very useful hung up near your bench. Apply polish to the string with a little lighter fuel or paraffin and pick up a few threads; thread through very small holes to clean and polish the interior.

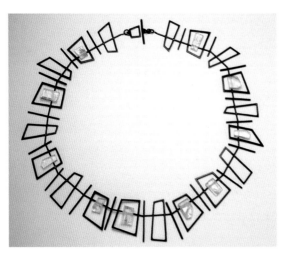

OXIDIZED FINISH

This silver necklace has been oxidized, to give it a distinctive black appearance. A solution containing ethanol and ether was painted onto the warm metal, which was then waxed. Colored stones provide a complete contrast to the black metal.

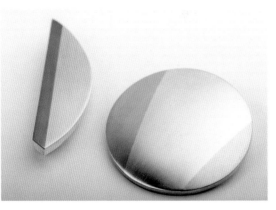

TWO-TONE MATTE FINISH

These brooches were given a matte finish, which makes the contrasting metals sit happily together. Any firescale would have been removed before their final finish with wet and dry papers and pumice powder.

Lesson 42: Hand finishing a plain band

A really good finish can be achieved by hand with the correct use of files, wet and dry papers, and polishing cloths. There are many different items on the market for finishing. Some are for specific metals such as platinum, and you will eventually find what method suits you.

1 The silver band has been soldered together. File away any excess solder using a flat file on the outside and an oval file on the inside.

2 File until you can no longer see the solder join. Be careful not to file away any thickness from the ring itself because this will spoil the overall look when it is finished.

3 Place the ring on the bench pin. Use a flat file all around the outside. You will find that the edges will pick up the file first, so continue filing until the surface is completely smoothed by the file. Then use an oval file around the inside.

4 With some 240-grit wet and dry paper, stuck onto a flat piece of wood to form an "emery stick," work around the outside and inside to remove the scratches left by the fine files. Continue with the wet and dry papers, working through 400, 600, and 1200 until the surfaces are smooth and shiny.

5 Now use either a polishing cloth or a liquid cleaner to really polish up the metal. Fasten a duster in the bench vise and pull it tight. Thread the ring over the cloth and rub hard along it.

 PRACTICE PROJECTS

Use this technique to make the cabochon-set ring; see page 126.

Lesson 43: Machine polishing a plain band

A polishing wheel is an expensive item to start with, and there are other ways of achieving a good finish, so don't rush into buying one. The first polish with a machine is the coarsest, and the last polish gives a final high shine.

1   Before starting to polish, follow steps 1–4 of Lesson 42. Hold the arm of the polishing wheel and screw the muslin wheel onto the end of the shaft. If it is a new wheel, follow the instructions on the right before use. Think of the buffing wheel on the spindle as a clock: 12 o'clock is at the top. Hold the work in two hands between 4 and 5 o'clock (3 o'clock should be facing you and 6 o'clock is at the bottom).

2   Charge the buffing wheel with polish by holding the bar between 4 and 5 o'clock and pushing it down and away from you.

3   When you polish a ring, you are holding one circle against another so they meet in one place. Therefore you must turn the ring around so that each area makes contact with the buffing wheel. The edges of the ring are polished by holding it across the wheel to give the edges direct contact with it.

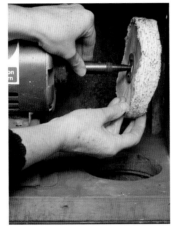

PREPARING THE MUSLIN BUFFING WHEEL

Prepare the buffing wheel by burning off any stray pieces of cotton with a small lighter. Next, fasten it onto the shaft of the polishing wheel and switch this on. As it is turning, hold the metal-end handle of a file horizontally against it to remove any further loose ends.

POLISHING MACHINE SAFETY

The polishing machine can be dangerous, so it is essential to make sure that:
- you have no loose sleeves, long hair, chains, or bracelets that could get caught up around the spindle;
- you know how to use the stop/start button;
- you are not distracted while using the machine.

Always wear eye protection, and make sure the machine has a dust collector attached to it to catch the fine polishing particles.

Unit 25: Dapping and swaging

Try shaping metal into useful and interesting shapes in dapping and swaging blocks. A small perfectly formed dome works as a base for many items of jewelry. Domes can be soldered together to make spheres and always look good with a little stone set in the middle.

SEE ALSO
Unit 15: Annealing, page 58
Unit 23: Using stakes, page 80
Unit 29: Hammering, page 101

DAPPING AND SWAGING TIPS

Dapping punches and blocks: Keep punches clean and unmarked, and check dapping and swaging blocks to make sure no foreign bodies are in the bottom of the curves, which would mark any metal being shaped.

Texture: Texture metal before dapping or swaging. It is impossible to add texture after the forming has taken place. When forming a textured piece, use a wooden punch because it will not mark or flatten the pattern of the texture.

Size: A circle of metal to be dapped should not be larger than the largest hole in the block. If you try to squeeze a larger piece of metal into a smaller curve, it will mark around the sides where it comes into contact with the edges in the block.

Mallets: The top of a metal punch can be hit with a steel hammer, but when using the side of a punch in a swaging block, only hit it with a wooden mallet.

Annealing: Metal up to about $\frac{1}{8}$ in (3 mm) thick will form quite easily in the dapping block but only if it is annealed properly.

A set of steel punches and a dapping block can be expensive, so to begin with, choose a set of wooden punches and a brass block because they will do most of what you want very satisfactorily. Other kinds of punches and blocks will form oval, triangular, and rectangular shapes, all of which will make interesting additions to your designs.

Lesson 44: Making a dome

It is possible to raise domed and curved shapes by hand with just a mallet and a stake, but a dapping block and punch make it a much neater and quicker job.

1   Mark out a $\frac{3}{4}$-in (15-mm) diameter circle on some 18-gauge (1-mm) metal sheet with a plastic template and a scribe. Only go around the circle template once, to leave a single cutting line.

2   Pierce out the circle, keeping the blade just to the outside of the scribed line.

3   Anneal, quench, and pickle the circle. Dry it well and then place it in the dapping block, inside a curve that is slightly bigger than its circumference.

4   Find the metal or wooden punch that fits into the curve you have chosen. Place the round end of it in the center of your metal circle.

5   With a wooden mallet or a flat-headed hammer, hit the top of the punch sharply to push the circle down into the curve. Use the punch with the hammer until you have an even dome shape.

DAPPING TIP

To work a dome up into more of a half sphere, place the dome in a smaller curve and with a smaller punch, hammer it again. Work the punch from the sides down to the center until the dome starts to close up. You may have to continue placing it in smaller curves until you get the shape you want, annealing in between if necessary.

Lesson 45: Making tubing

A swage block has different shaped indentations with punches to match. One simple type of swage block has half-round channels, which can be used with the round handles of punches to start making tubing. Other types of swage blocks have rectangular, triangular, and square shapes.

1   Cut a strip of silver sheet and anneal it. Find the curve in the block that the strip of metal will just lay into. Place a round former along the top of the strip and use a wooden mallet to hammer it down into the curve.

2   Now place the metal into the next curve down and repeat the process, annealing in between as necessary. The formers will need to become smaller as the curves decrease.

3   As the metal starts to close up around the former, remove it from the block and place on a flat surface. Gently hammer the metal over the former to close it. You may need to use a slightly smaller former to close the seam. You can now solder the join in the tubing (see page 74).

Unit 26: Drilling

There are several different methods of drilling holes in metal. Usually, in jewelry making, the holes are quite small—under the size of a #46 (2-mm) drill bit—and the smaller the drill bit is, the more expensive it is. Define a location point before starting to drill to prevent the drill bit from wandering, which would mark the surface of the metal and risk it breaking.

DRILLING TIPS

Keep spare bits: Buy more than one drill bit of the size you need. There's nothing more annoying than breaking the only one of the right size.

Flatten curves: On a curved surface, use a flat file to make a little flattened area before marking the location point.

Open holes gradually: Use lubrication such as oil on the drill bit. This will keep it sharp and prevent it from overheating. Use a small drill to start with and gradually increase the size until you reach the right diameter.

Hold the drill upright: Hold the drill at a 90° angle to the metal.

Remove the metal shavings: When drilling a deep hole, turn the drill in the other direction every so often to bring out the metal shavings. Buildup of metal may cause the drill bit to break.

Remove broken drill bits: If the drill does break inside the metal, drill another hole close by so you can reach the broken drill bit with either a saw or a sharp point to remove it. If it is not removed it will cause problems when pickling and will devalue the piece.

Countersink holes: To soften the sharp edges of drilled holes, take a larger drill bit and turn it gently with your hand to countersink the hole.

DRILLED STONES
The stones on this bracelet were drilled with a stone-cutting diamond burr.

Types of drills

Several types of drills are available, each of which has a specific purpose.

SMALL HAND DRILL
A small hand drill is used to drill tiny holes. It can be turned quickly or slowly. The drill bit should be tightened into the chuck almost up to the cutting edge. Hold it upright at 90° to the work with the top tucked into the palm of your hand, and use a gentle downward pressure as you turn it through the metal.

SEE ALSO
Unit 14: Piercing, page 52

ARCHIMEDEAN DRILL

An archimedean drill is a handheld drill that works by a smooth up-and-down movement of a horizontal bar. Hold the bar in one hand, twisted around until it is at the top of the drill. Allow the bar to fall and rise, to drill a hole in the piece of metal.

WOOD DRILL

A wood drill is useful for drilling slightly larger holes. Because it requires both hands, the metal or wood must be made fast, in a vise or taped down onto a wooden surface. One of its main uses in the jewelry workshop is to twist wire, as shown here.

FLEXSHAFT MOTOR

A flexshaft motor can be used for all sorts of applications. For drilling, fasten a drill bit of the correct size into the chuck and hold the shaft absolutely upright in one hand while the other holds the work steady. Make sure you are holding the work on a soft surface, such as wood, so the drill is not harmed as it goes through the metal.

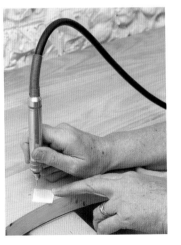

Lesson 46: Using a small hand drill

Always begin with a small drill bit and gradually open the hole up to the larger size — in this case, ⅛ in (3 mm). Working in this way, the location is much easier and the opening will be truer. Using the correct size drill bit right at the beginning tends to make the hole a fraction too big.

1   Tap a sharp pointed punch with a ball-peen hammer where you would like the drill hole. Do this over a steel block or plate. Fasten a #56 (1-mm) drill bit into the chuck of the hand drill. This should be tight so it does not slip as it is being turned. If necessary, fasten the handle of the drill in the safe jaws of the vise and tighten the end with a pair of pliers.

2   Hold the top of the drill in the palm of your hand and start to turn it. Keep it at right angles to the work and apply a little pressure to make sure the drill can start to bite. Too much pressure can cause the drill bit to break. Continue until you have drilled right through, bringing it out every so often to clear the shavings. Next, fit the #46 (2-mm) drill bit and repeat the process. Continue increasing the bit size until you have a #30 (3-mm) hole.

SEE ALSO
Unit 20: Joining, page 70

Unit 27: Fittings

Fittings are the means of attaching or closing jewelry so it sits comfortably on the wearer. There are many established fitting designs, but it is always interesting to try to create a method of fixing unique to your piece. Test out new ideas to find the best solution. Many types of fittings can be bought from a metal supplier, and then the fitting design is just a question of placement.

When designing the fittings for a piece of jewelry, ask yourself three questions. First, "What is the fitting for?" For example, it could be a closure or a pin of some kind. Second, "Where will the fitting be placed?" It might be concealed or turned into a feature. Finally, "Does the weight of the fitting match the weight of the piece?" It must be strong enough to hold but should also look appropriate.

FITTINGS FILE

Necklaces:
box catch   trigger clasp   bolt ring
S-loop   hook   loop and T-bar   end cap

Chains:
bolt ring   trigger clasp   jump ring
loop and T-bar

Cuff links:
chain   T-bar   hinge fitting

Bracelets:
box catch   loop and T-bar
hinge and pin   S-hook

Earrings:
hooks   hoops   clips   posts

Pendants:
hook   loop   bail   tube

Brooches:
hinge and catch with pin
tube-in-tube   pin catch

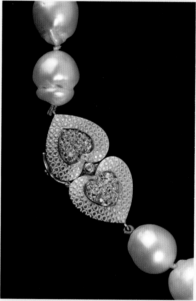

DETAILED BOX CATCH
Here the idea of a simple box catch has been turned into a beautiful centerpiece, even though it will be worn at the back of the neck.

MULTIWAY FITTING
This detailed fastening serves two purposes: it is a means of fastening the chain and it provides an interesting central decoration. The chain could also be worn long with the ring and bar hanging loose.

Lesson 47: Making an S-loop

An S-loop is a simple and effective way of closing a chain or a necklace. Use round wire that is appropriate to the size of the piece and has been annealed (see page 59).

1   Cut a length of wire longer than you need, or work from one end of a coil of wire. File the end to a point. Start it about ⅜ in (10 mm) from the end so it goes in gradually, and make sure the point is not too sharp.

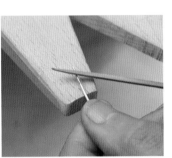

2   Use a pair of half-round pliers to make a slight curve outward on the point.

3   Use the half-round pliers to hold the wire about ¾ in (20 mm) from the point, and make the first part of the S. Take the wire down past the point. With a pair of dividers, measure the distance from the point to the first S. Mark that distance on the wire that goes past the point with a marker pen.

4   With the half-round pliers, make an equivalent bend at the other end from the first curve and bring the wire back past the point again.

5   Cut the wire so the two ends finish up opposite each other. File the second end to a point similar to the first. Make a small outward curve with the half-round pliers to match the first.

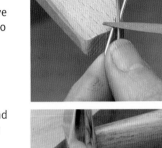

6   The first S fits onto one end of the chain and is usually soldered where the point touches. The other end is opened slightly so it will fit into the jump ring or hook at the other end of the chain. To make the S springy, gently hammer both ends of the S with a riveting hammer on a small anvil or flat plate after it has been soldered.

END-FINISHING TIP

Instead of filing the ends of the S-loop to points, try melting up the ends of the wire so they form small balls on the ends. (See page 123.)

Lesson 48: Applying commercial fittings

Many commercial fittings are supplied already assembled. Most fittings such as ear clips or cuff links (shown in this lesson) should be taken apart before soldering; otherwise the vital "spring" parts will be lost as heat is applied.

1   Remove the rivet that holds the cuff link fitting together. Prepare the back of the cuff link by filing it with a small flat file. Paint a little flux in the center and place a paillon of easy solder in it. (If the fitting has a broad back, you may need two or three paillons.) Solder the fitting onto the back as in Lesson 37, page 75.

2   If necessary, clean up around the soldered area. Place the T-bar onto the top of the fitting, lining up the holes.

3   Hold the small wire rivet in a pair of tapered flat pliers and push it carefully through the lined-up holes.

4   File the ends of the rivet flat on each side if necessary. Steady one end of the rivet on an anvil and tap the other end with a riveting hammer to spread it. Then turn it over to spread the other end.

Lesson 49: Making a brooch fitting

There are many different brooch fittings. This is just one of many, so experiment! Always ensure that the fitting keeps the brooch in place without it tipping over, that it won't ruin clothes, and that it isn't impossible to fasten.

1   First make the hinge. Cut a $\frac{1}{4}$ in x $\frac{1}{2}$ in (5 mm x 16 mm) piece of 20-gauge (0.8-mm) silver sheet. Mark lines $\frac{1}{16}$ in (1 mm) on either side of the center line. Saw a little into both lines; then use a triangular file to make a groove just over halfway through the silver.

2   With a scribe, mark the center point of one of the end sections. Use a hand drill fitted with a #56 (1-mm) drill bit to drill a hole through the section.

3   With a pair of flat-nosed pliers, bend each end section inward by 90°. Run a piece of hard solder up the outside of the corners to strengthen them (see page 74). This piece will form the joint.

4   Place the joint with the undrilled end section lying flat on the bench. Put the #56 (1-mm) drill bit through the hole and drill a second hole through the undrilled section of the joint. Solder the flat back of the hinge onto the back of the brooch.

5   Now make the catch. Use D-section 12-gauge (2-mm) silver wire. Cut a length about 1 in (30 mm) long. Use a pair of round/flat pliers to make a tight turn at one end.

6   Bring the curve right around, and then turn the rest of the length at a right angle.

7   Saw away some of the curve to allow for the pin.

8   Squeeze the hook a little with a pair of pliers and then cut off the long end of the wire (the base) to just a little longer than the hook. File the base completely flat. Hammer lightly with a riveting hammer on a steel surface. This will workharden the catch.

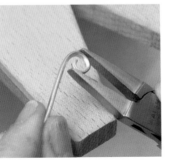

9   Make the pin with a length of 18-gauge (1-mm) round wire. With a pair of flat pliers, make a sharp turn in one end of the wire. Use the round/flat pliers to make three quarters of a circle next to the bend.

10 Bring the wire out from the circle in a straight line. Measure how long it needs to be by placing the circle in the hinge and then trimming the wire to a length where it will meet the catch. File the end to a blunt point.

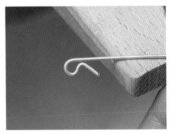

11 The pin is held in place with a rivet that fits neatly into one side of the hinge, then through the circle of the pin and out through the opposite hole of the hinge. The two ends are filed and riveted (see Lesson 54, page 102). When the hinge is attached to the back of the brooch, the bent end of the pin should press against the brooch back to provide the spring action.

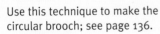

PRACTICE PROJECTS

Use this technique to make the circular brooch; see page 136.

BROOCH FITTING TIP

Once the brooch is formed, twist the wire of the pin on itself to strengthen it. Grip the wire near where it joins the brooch, and again halfway along, and give it a quarter turn. Lightly hammer the pin on a steel surface to workharden it.

SEE ALSO
Unit 9: Precious metals, page 34
Unit 10: Nonprecious metals, page 38

Unit 28: Casting

Metal is cast by melting it and pouring it into a mold. Casting is used to make three-dimensional pieces that would be very difficult to make in any other way. It is also used to make more than one item of exactly the same design. The better the precision of the mold, the better the finished casting will be.

QUANTITY OF METAL

Sprue and button: Any casting must have a sprue—the channel through which the molten metal is poured—and a button—the opening at the top of the sprue where the first pour is made. Sprues and buttons are cut away and can be reused after pickling and cleaning. When estimating the weight of metal needed for a casting, the weight of the sprues and button should always be taken into account.

Wax to silver: To calculate the amount of silver needed to cast from a wax model, multiply the weight of the wax by 11. Add 3 dwt (10 g) of silver to allow for the button.

Displacement method: To calculate the amount of silver required for a model not made from wax, use the displacement method. Fill a measuring jug halfway with water. Place the item to be cast in it and note the new measurement. Remove the item. Place casting silver grains or scrap silver in the water until it reaches the right level. This will be the correct weight of silver. Add 3 dwt (10 g) for the button.

Take solidity into account: If you use anything hollow like a metal bead or a shell to make an impression for a casting, don't forget that the finished piece will be solid and therefore weigh much more than the original.

There are three main methods of making a mold. One method is "cuttlefish casting." A dried cuttlefish bone is sliced in half lengthwise, and either a model is pushed into it to make an impression or the bone is carved out. The two halves are placed back together, and molten metal is then poured into the cavity made in the cuttlebone.

The second method is "sand casting." With this method, an impression is made in the sand by a model, which is then removed, leaving a cavity. Molten metal is poured into the cavity to form a replica of the model.

There is also the "lost wax" method. A wax model is made, then surrounded by a powder mix called investment. This is left to dry and then placed in a kiln. The wax melts out of the investment, leaving a cavity that can then be filled by the molten metal.

When deciding whether to use a casting method, always ask yourself, "Could I fabricate this piece with metal sheet or wire?" If the answer is no, then go ahead and cast it.

CAST RINGS
Here, the artist used casting to make two rings based on a marine theme: a giant clamshell ring with pierced holes, and a stacking ring utilizing applied textures and waves on separate bands.

Lesson 50: Cuttlefish casting

The dense backbone of the cuttlefish makes an ideal mold for casting with silver or gold. Use this method to cast a long length of silver with the beautiful pattern of the cuttlefish, which can later be annealed and shaped into an open bangle. This method of casting can only be used once, whether or not it is successful.

1   Use a saw frame to remove the pointed top and bottom of the cuttlebone. Now cut down the sides to remove the hard edges.

2   Stand the flat bottom on the bench and hold it steady. Place the saw frame straight across the top and gently push the blade down through the bone to cut it in half lengthwise.

3   Take the thickest half and start carving away the bone with a carving tool, trying to keep to a consistent depth of about ¹⁄₁₆ in (2 mm). Carve out an area the length of the finished bracelet.

4   When you have finished carving, dust the whole area with a dry paintbrush. This will give the natural texture of the cuttlebone a very clear definition.

5   With a pencil, draw arcs from the top of the shape to the top of the bone, for the button. Scoop out the area with a carving tool. Make a matching scoop on the other half of the bone. Draw some air lines out from the bracelet shape.

6   Place the two halves back together, making sure that they are aligned. Fasten them together at the top and bottom with binding wire, twisted tightly.

7   Hold the cuttlefish firmly on the soldering tray, so it cannot slip or fall over while the metal is being poured into the top opening. Using the cuttlefish in this way gives the item being cast a flat back.

CUTTLEFISH CASTING TIP

You can buy cuttlebones from your local pet store. The best bones to choose for casting are thick and as large as possible. Thin ones, which "dip" inward, are more likely to break when held together for pouring.

PRACTICE PROJECTS

Use this technique to make the cast cuff links; see page 129.

CASTING A PIECE WITH A SHAPED BACK

To achieve a shaped back, rather than a flat back (as shown on page 97), the other half of the cuttlebone is carved with the same outside profile as the first. With this method it is very important that the two sides fit exactly when they are placed back together. After the cuttlefish is cut in two, rub each side on a piece of 240-grit emery paper until they are completely flat, to make sure everything will line up when you place the halves back together.

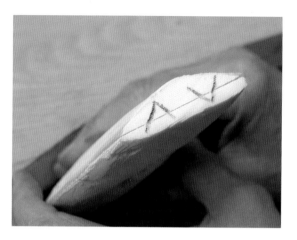

LINING UP THE HALVES
Try scribing a couple of lines across the flat top face of the fish before it is halved. These lines can then be lined up when the two halves go back together.

Lesson 51: Sand casting

The sand-casting method uses the "Delft" system. This consists of two aluminum rings that slot together to form a tube. The rings are packed with "casting sand," which is a very dense, slightly oily sand. The sand cannot be carved, so this method of casting can only reproduce an object that already exists. Of course you could make this object yourself, in wood, wax, acrylic, or anything else that will withstand a certain amount of pressure as it is pushed into the sand.

1 Divide the aluminum rings. Place the shorter one on a sheet of paper on the bench. The connecting lip should be uppermost.

2 Pack the short ring with the sand, pressing it firmly down into the ring until it is slightly higher than the top. Level it off with the edge of a ruler.

3 Press the model into the sand face down until about half its thickness is buried. It may need gentle tapping with a mallet to help it sit down into the sand. (Be sure your model is not going to crack or break if you do this.)

4 Leave the model where it is and sprinkle a little baby powder over the surface of the sand. This will help the two halves to separate later. Spread the powder evenly with a paintbrush.

5 Fix the taller ring back onto the short one, lining up the two notches marked in the sides. Now fill the top ring with sand. When it is full, use a mallet to hammer the sand down so it is as dense as possible.

6 Divide the two rings again. Remove the model from the bottom ring very carefully. You should have a cavity the shape of the model in the sand on each side of the ring. Draw some air lines out from the lower cavity.

7 Make a hole for the sprue in the top, larger half, with a #46–#30 drill bit (2 mm–3 mm diameter). Twist the drill up from the top of the cavity made by the model until it breaks through the top of the sand.

8 Where the drill has broken through, open out the area to make space for the button. If necessary, clear any sand from the sprue channel. Now you can put the two halves back together again, making sure the notches are lined up, and place the mold onto the soldering tray ready for casting.

SAND-CASTING TIPS

Experiment with sprueing: Sand casting relies on gravity up to a certain point. Once the metal is poured down the channel, it then has to spread out to fill the cavity. Sometimes it will take a few tries to get it right or indeed to fill everything. Experiment with the sprueing; it may need to be thicker, or thinner, or you may need a larger button. Make sure the air lines are clear.

Casting hollow objects: Any model that is 1/16-in (2-mm) thick or less will not cast well. If you want to cast something like a shell that is hollow, fill it with wax, or the sand will try to fill it and the two halves will not separate cleanly.

Lesson 52: Melting and pouring silver

Whichever casting method you use, a successful result depends on the techniques you use to melt and pour the metal—in this case, silver—into the mold.

1 Cut and weigh out the necessary amount of silver and add 3 dwt (10 g) for the button. Place it in a small crucible that is held in a handheld grip. Sprinkle a little boric powdered flux over the silver. Use as large a flame as possible to heat the silver.

2 Keep the hottest part of the flame over the metal until it glows bright orange. Keep it there until the metal starts to shine and melt. If the flame is too far away the metal will start to solidify and oxidize.

3 Once the metal is melted it will form a ball, which will start to spin. Let it spin without altering the flame while you count slowly to 20. It should be ready to pour. Gently bring it over to the opening in the cast. Tilt the crucible so the molten metal is right at the edge. Pour it down the hole with one smooth movement. Wear leather or insulated gloves to open up the mold. Drop the hot silver model into some cold water, and then pickle.

POURING TIPS

Go with the flow: Try to imagine how the metal will flow when it is poured into the mold. It is either pulled by gravity or spreads out where it can. The one thing it won't do is go back on itself.

Keep casting metals clean: Save clean scraps of metal to use for casting. Make sure no solder is left on any of the metal because this will degrade the metal being melted. Anything suspect should be cut or filed away.

PRACTICE PROJECTS

Use this technique to make the cast cuff links; see page 129.

SAFETY TIP

Take care when working with molten metal. Make sure your cuttlefish or casting system is held within an edged soldering tray so any spillages will fall into that and not onto the surrounding area.

Unit 29: Hammering

Jewelers use different weight hammers for a variety of purposes. Steel hammers are used to stretch, spread, or texture metal. They can also be used with a punch or chasing tool that shapes and marks the metal. When a steel hammer strikes directly onto metals such as copper, silver, or gold, it always leaves an imprint.

A medium-weight ball-peen hammer can be used for most applications, but a small, light riveting hammer is invaluable for more delicate work. Large, heavy hammers are mainly used for texturing; the rougher the face of the hammer, the more interesting the texture will be. Larger hammers with smooth, clean heads are used to "planish," or smooth out, metal that has been previously hammered. Leather rawhide, wooden, or plastic mallets are used to shape the metal without marking it (except very soft metals such as fine silver and gold); generally they can be used with some vigor without leaving marks.

HAMMERED EDGES
Hammered textures at the edges contrast with the galena centerpiece in this highly original brooch.

Lesson 53: Stretching metal with a hammer

You can use a hammer to stretch a length of 6-gauge (4-mm) round wire so it comes to a pointed end of around 12-gauge (2 mm). Using this method, no weight of metal is lost; if the metal were filed down from 6- to 12-gauge (4–2 mm), quite a lot would be wasted.

1   Anneal a 4-in (100-mm) length of 6-gauge (4-mm) round silver wire. Quench, pickle, and dry it. Hold one end of the wire on an anvil. Begin hammering the silver about 1¹⁄₂ in (40 mm) from the end. Hammer down to the end, flattening the side.

2   Give the wire a quarter turn and hammer it down in the same way. As soon as the wire starts to feel hard, anneal it. Continue to work like this until the end has tapered down to just over 12-gauge (2 mm) thick.

3   Lay the wire across the anvil. Roll the wire along the length of the anvil with one hand and hammer smoothly with the other, using small, even strokes to make the rather square wire become round. After stretching the metal like this it will require a little filing to smooth it down. It can be finished with wet and dry papers.

Lesson 54: Making a rivet

A rivet is a small piece of wire that fits through two or three thicknesses of metal to hold them together. The ends of the wire are made thicker, so they are larger than the holes and cannot fall through them.

1 Place the pieces of metal together, so the holes in them are lined up. The wire for the rivet should be the same diameter as the holes. Use a longer length of wire than you need, and push it all the way through the metal pieces until there is about $\frac{1}{32}$ in (1 mm) protruding on the other side.

2 File the protruding end so it is completely flat against the metal. Use a pair of top cutters to snip off the other end of the wire to within about $\frac{1}{32}$ in (1 mm) of the metal, and file that end flat.

3 Carefully lift the whole ensemble onto a small anvil. You may need a friend to supply a third hand for the next part. Place the pointed end of a small center punch in the center of the rivet, and tap it carefully with a hammer. Turn the pieces over and do the same to the other end of the rivet. Now use a small center punch with a broader head and tap all around the end of the rivet. This will spread the end of the rivet out to look somewhat like a nail head.

4 Turn the pieces over again, and spread the other end of the rivet in the same way. File any sharpness away and finish with some smooth wet and dry paper.

SEE ALSO
Unit 15: Annealing, page 58
Unit 31: Using a rolling mill, page 106
Unit 33: Texturing, page 115

Lesson 55: Texturing with a hammer

When using a hammer to texture metal, remember that the metal will stretch considerably. It is better to texture first and then cut the shape.

HAMMERING TIPS

Creating an overall texture: The rounded end on any hammer can be used to create an overall texture of small indentations in the metal. The heavier the hammer, the deeper the indentations will be.

Using the wedge: The wedge-shaped end of a hammer can also be used to texture metal. The smaller the hammer, the slimmer the line of the texture will be.

Test the effect: Try any new texturing methods on copper or scrap metal to see if you like them first.

Care for your hammers: Keep the heads of hammers clean and shiny if they are being used to stretch or shape metal rather than texture or mark it. First clean them with wet and dry paper, and then polish them with white diamond polish on the polishing machine.

Using mallets: Wooden, plastic, or leather rawhide mallets are softer than metal hammers and will become damaged after a time. As long as they still work, don't worry! Use them to round up rings and bangles on the mandrel and to shape metal around formers.

1 Anneal a small piece of copper sheet. Hold it on a metal flat plate and use the "ball" end of a small ball-peen hammer to make little round marks all over the piece. Try to keep the marks close together to produce a uniform texture.

2 Anneal the piece again, and use a plastic template to draw two circles on the smooth back of the metal. Pierce out the circles. These could now be domed using the wooden punch in the dapping block (see page 88).

USING A CHASING HAMMER

Chasing is marking the metal in lines and curves to delineate a shape or scene. The work has usually been worked previously from the back so when it is turned over it is raised. The work is then held firm with a substance called "pitch," which is poured into the back and allowed to set. It is then turned over and held in a bowl or wooden lock and worked from the front with "chasing tools" and a chasing hammer. The term "repoussé" refers to working the metal from the back.

The chasing tools are held at a slight angle to the line they are making, with the end nearest to you and the top away from you. The top is hit gently and consistently with the hammer so the tool is pushed smoothly along the line.

Unit 30: Using draw plates

Draw plates are used to reduce the diameter of wire or to produce a different profile. A round-holed draw plate is very useful for the times when you need a piece of wire of a very specific size, for example to make rivets (see page 102). Wire profiles such as triangular, square, rectangular, and oval can all be made using an appropriate draw plate. You can buy these different shapes from your metal dealer, but usually they come in quite long lengths and are a lot more expensive than silver wire you can draw down yourself.

WIRE-PULLING TIPS

Anneal regularly: Keep the wire you are pulling down well annealed. You will feel it become more difficult to pull as it hardens.

Maintain the taper: Keep the end you are pulling through filed to a taper. The hole at the front of the draw plate is narrower than the hole in the back. The end of the wire must protrude far enough at the front to be gripped with a pair of pliers or draw tongs.

Reduce diameter gradually: Pull down one hole at a time. For example, if you are pulling 12-gauge (2-mm) wire down to 14-gauge (1.6 mm), the first hole to pull through is 13-gauge (1.8 mm). Pull the wire down through smaller holes until you reach the required measurement.

Use safe jaws: Safe jaws must be used to hold the draw plate; serrated jaws would close over some of the holes in the plate and damage them. Safe jaws are made by simply bending two strips of aluminum or copper into right angles and placing them on the jaws of the vise.

Anticipate the end of the wire: You will need to pull quite hard on the wire, so take care when it is almost pulled through to avoid flying backward across the workshop!

Lesson 56: Pulling down round wire

Use the draw plate to pull down some round wire from a diameter of 18-gauge (1 mm) to 24-gauge (0.5 mm). Decide on the length you want; don't forget that the length will increase as you pull the wire down.

1   Cut a length of 18-gauge (1-mm) round wire. Anneal it, then pickle, rinse, and dry it thoroughly. Use a half-round file to make a long, tapered end (see page 77).

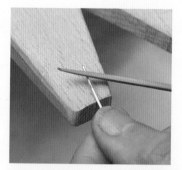

2   Fix the round draw plate into the safe jaws of the vise. Keep it straight and make sure the holes you will be using are clearly visible.

3   Melt a few drops of beeswax onto the wire behind the tapered end, or use a drop of oil on the hole that the wire is to be pulled through.

4   Push the tapered end through the 18-gauge (1-mm) hole. There should be at least ¹/₂ in (10 mm) protruding. If there is less, file more from the taper until you can push more of the wire through.

5   Grip the tapered end of the wire in a pair of serrated-edge pliers and pull it, keeping it straight. Once you have pulled it through the hole, repeat the process with the 20-gauge (0.8-mm) hole.

6   After drawing the wire through the 22-gauge (0.6-mm) hole, you may wish to anneal it. Wind it into a circle and play a gentle flame around it. Turn it over to anneal the other side. Quench, pickle, and dry. File the taper and pull through the plate until you reach 24-gauge (0.5-mm) diameter.

Lesson 57: Pulling square wire down into round

When altering the profile of the wire, the finished shape will be smaller than the widest dimension of the original wire. Use a length of 12-gauge (2-mm) square wire.

File the end of the square wire into a taper. Place it through the 12-gauge (2-mm) hole in the round draw plate and pull through. Anneal the wire; then pickle, rinse, and dry it. Place the tapered end in the nearest smaller hole, place some beeswax or a drop of oil onto the wire behind the plate, and pull it through. Continue pulling down the square wire through smaller round holes, annealing as necessary, until it has become completely round.

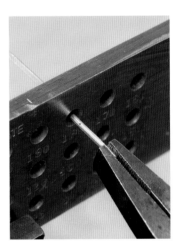

SEE ALSO
Unit 15: Annealing, page 58
Unit 34: Wirework, page 119

Unit 31: Using a rolling mill

A rolling mill is a very useful piece of equipment, primarily used to decrease the thickness of metal. The metal sheet is placed between the two rollers and a handle is turned until it comes out the other side; the space between the rollers is adjusted easily. A rolling mill can be used for texturing sheets of metal (see page 116), and the bigger ones can also be used for shaping wire.

Metal that is softer than the steel of the rollers will decrease in thickness when pressure is applied through it. Metal that is harder than the steel will mark the rollers. So never use binding wire, titanium, hard steel, or anything else that you think is harder than the rollers, in the mill. Metal with uneven edges will distort as it is rolled through the mill. This usually does not matter as it can be cut afterward, but if you require your piece to stay exactly the same as it is being rolled, make sure the sides are parallel.

Buy the best rolling mill you can afford. Be careful if you buy a used rolling mill; check that the rollers are in good condition, that they are not dirty, rusty, or pitted and they sit together evenly. Rollers can be professionally reground to smooth them down—but check that you can get this done before committing to an old machine.

SMALL ROLLING MILL
Smaller mills—probably the most practical for a small workshop—only have one pair of smooth rollers. A large rolling mill has another set of rollers underneath the smooth top pair. These have a ridged or V profile and are used to roll down large diameter wire or to roll down an ingot that has previously been melted and poured into an ingot mold.

ROLLING MILL TIPS

Use dry metal: Never put wet or damp metal through the rollers. Always use a paper towel to wipe metal before putting it through the mill. Any dampness will leave a rusty mark on the rollers, which in turn will cause pitting.

Roll down gradually: If the metal is subjected to too much stress when being rolled down it will crack when next annealed. Don't try to roll down too much too soon; just a little bit each time is a safer way to go.

Oil the mill: Near the top of the rollers there is an oil hole. Remember to squeeze a little oil into it from time to time to prevent the cogs from going rusty and to keep the whole mechanism in good condition.

Remove marks: If any marks do occur on the rollers, first wipe them over with a cloth dampened with acetone to remove any dirt. Any further marks can be carefully removed by rubbing over them with some fine wet and dry paper (used dry). Fine steel wool (4/0) will also remove any debris on the rollers.

Keep the mill covered: It is a good idea to cover the mill when it is not in use. It will help to keep it dry and to protect the rollers from workshop debris.

It is best not to use bottled gas to heat your workshop. Gas creates a lot of condensation, which settles on anything cold and metallic, making objects rusty before you know it. Radiant or electric heat is much better.

Lesson 58: Rolling down silver

In this lesson you will roll down a 2-in (50-mm) square of 16-gauge (1.2-mm) thick silver to a thickness of 22-gauge (0.6 mm). This means it will double in size.

1   Cut the silver to the above dimensions and anneal it. Quench, pickle, and dry it thoroughly. Open up the rollers, place the edge of the metal on the bottom roller, and wind the dial around until the top roller just catches the top of the silver. Wind the handle of the mill and roll the metal piece all the way through.

2   Catch the silver as it comes through, and then turn it around a quarter so that what was the side edge in the first pass becomes the leading edge in the second.

3   Wind the rollers down again, move them roughly about 5 points, as shown on the mill dial, and pass the metal through again. When the rollers have been wound down 15 points, in other words three times, anneal the silver again to prevent any stress building up. Continue in this way until the thickness of the metal measures 22-gauge (0.6 mm).

Unit 32: Stone setting

Traditionally, one of the main purposes of a piece of jewelry was to show off a beautiful stone. Stones can be cut and shaped in many different ways, some of which are described in Unit 11 (page 40). This unit deals with the metalwork involved in making the "mount" for a stone to sit in and how this best attaches to the main piece.

The examples shown here illustrate some of the many possible ways of setting stones.

SMOOTH CURVES
In this dramatic ring, the rub-over setting for the iridescent fire opal forms a continuation of the smooth curves of the ring shank.

TEXTURED BASE
In this pair of stud earrings, the jeweler has imprinted a circular pattern to add interest to the base of the rub-over settings for these lovely moonstones.

MULTILAYERED RING
In this stacking ring, both channel and rub-over settings have been used. All the settings are different, which adds visual interest, and being mounted on separate bands permits the combination to be rearranged.

JEWELED BRACELETS
A collection of differently shaped and sized cabochon stones are held by simple rub-over or pin settings. The designs achieve unity through the inclusion of the silver rings and wooden beads threaded onto the bracelets.

SEE ALSO
Unit 11: Stones and beads, page 40

Lesson 59: Setting a round, flat-backed cabochon

It is a good idea to start by setting a cabochon stone. A stone with a flat bottom sits comfortably in a flat-bottom setting; if it is at all curved it will wobble, so allowances must be made for this. The setting shown here is for a 3/8-in (8-mm) flat-backed cabochon stone designed to be soldered onto a pendant.

1   Look to see how the sides of the stone slope in. The point where the stone starts to slope is the height that the bezel needs to be, so when it is pushed over onto the stone it will hold it in.

2   On a piece of 24-gauge (0.5-mm) silver sheet, mark the length of the bezel —in this case, around 1 in (27 mm). See calculating the size of a bezel, page 111. Open the dividers to the height of the bezel, and draw a line along the length.

3   Cut out the bezel, then anneal, quench, pickle, and rinse it. When it is dry, use a pair of half-round pliers to bend each end of the strip around so they meet exactly.

4   Solder the join with hard solder (see page 74), pickle, and dry it. File away excess solder and place the bezel on a small round mandrel to round it up.

5   Make sure the bottom of the bezel is flat by rubbing it up and down with a flat file.

6   Position the bezel on the pendant. Flux around the base and place several paillons of medium solder inside it, ensuring that they are in contact with the bezel and the base. Solder the two pieces together; direct the heat onto the larger area of silver before bringing it near the bezel.

BEZEL-MAKING TIP

When making a bezel for a ring, place it on a base sheet as for a pendant, but apply the solder around the outside of the bezel. Heat the base piece, and after the solder has run, quench and pickle it. Saw away the excess metal outside the bezel, and file the outside edge until the join line has disappeared.

Lesson 60: Making a raised setting for a stone

This setting includes an inner bezel for the flat base of a cabochon to sit on, which supports a cabochon stone or the girdle of a faceted stone. This setting is used when the stone needs to be raised above the ring shank or piece to which it is being attached.

1   Find the length of the bezel required (see opposite). Allow enough metal in the height for an arc to be filed to fit the curve of the ring shank. If the setting is for a faceted stone, the height of the bezel should be just more than the height of the stone from the tip of the culet to the top of the table (see opposite). Cut a strip of silver to the size you have calculated. Anneal, quench, and pickle the strip before bending it up and soldering the join with hard solder.

3   The stone should now be able to sit level in the bezel, with enough metal around the outside to be pushed down onto it. (A faceted stone can be placed into the bezel so the girdle is supported by the inner one.) File the base of the bezel with a half-round or oval file until the shape fits the ring shank. Solder the bezel to the shank (see page 75).

2   Now make a smaller ring for the inner bezel. This should form a shelf that comes below the top of the outer bezel, allowing enough height to set the stone. Anneal the piece for the inner bezel, and bend it to fit neatly inside the outer bezel. Make any adjustments as necessary; then solder it in. The bottom of the two pieces should be level.

PRACTICE PROJECTS

Use this technique to make the cabochon-set ring; see page 126.

CALCULATING THE SIZE OF A BEZEL

You need to know the measurements of a stone before you can set it. For a round stone you need to know the diameter, for an oval stone the length and width, and for rectangular or triangular stones you will need to know the measurement of each side.

Here are some simple formulas to find the length of metal needed for the bezel.

Round stone: (diameter of stone x π) + twice the thickness of metal

Oval stone: $\left(\dfrac{\text{length + width of stone}}{2}\right)$ x π + twice the thickness of metal

Square, rectangular, and triangular stones: the sum of the length of each side

Always cut on the generous side because a bezel that is slightly big is much easier to deal with than one that is slightly small.

Faceted stones: A faceted stone has angled faces and will have a greater depth than a cabochon of a similar size. The bottom, rather than being flat, comes down to a point called the culet. The widest part of the stone is called the girdle, which is the area that must be set into a bezel. From the girdle there are angled faces that continue up to the flat top, known as the table.

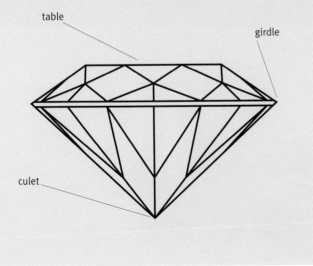

table

girdle

culet

Lesson 61: Setting the stones

To set a stone, the work must be held firmly so there is a resistance to the pushing that takes place as it is being set. Stone setting is always the last job to be done when making a piece of jewelry, so take care not to mark or scratch the piece in the process.

1 Position the ring in a clamp and hold it in the cutaway area of the bench pin.

2 If the metal for the bezel is thin—24-gauge (0.5 mm) or less—it will be comparatively easy to push it over onto the stone using a small pusher.

3 If the bezel is thicker, try using a small metal punch held at a 45° angle to the top of the bezel and tapped gently with a riveting hammer. To keep the clamp still, hold it in the safe jaws of the vise.

4   Clean the outside edge of the bezel with a fine needlefile and then with some fine wet and dry paper, being careful not to touch the stone.

5   Finish off the setting with a polished steel or agate burnisher.

OPAL-SET PENDANT
An opal nestles at the center of this striking gold pendant.

STONE-SETTING TIP

Pushing the metal onto the stone: Start by pushing the metal down onto the stone in one area, and then turn the work around so you can push the metal opposite the first area down onto the stone. Keep working from opposite side to opposite side until the stone is held fast, and then keep working all the way around until the bezel is smooth and there are no gaps between the metal and the stone.

Cleaning the bezel: When cleaning a bezel after setting a stone, be very careful not to touch the stone with the file or wet and dry papers. Try putting your thumb right over the stone to keep the file from drifting onto it.

TEXTURED RINGS
These simple rub-over settings contrast with the textured circular patterns on the wide ring shanks.

Lesson 62: Making a prong setting

Most faceted stones look better with light coming from behind as well as in front of them, and settings are designed to allow this. An open setting follows the taper of the stone from girdle to culet so the stone sits evenly and snugly in it. In this lesson you will use "claws" to hold a faceted rectangular aquamarine in an open tapered setting. The stone's measurements are ³/₈ in x ⁵/₁₆ in x ³/₁₆ in (10 mm x 6 mm x 5 mm).

1   Cut two 3¹/₈-in (80-mm) strips of half-round wire with dimensions of 14 x 14 gauge (1.5 x 1.5 mm) or slightly less. In the center flat area of one of the strips, use a half-round file to file down an area so the other length will fit into it. Solder the two pieces together with hard solder.

2   Bend 18-gauge (1-mm) round wire into a rectangle with outside measurements of ¼ in x ⅛ in (6 mm x 4 mm), with the join halfway down the longest side. Solder together with hard solder. Run a tiny paillon down each corner to strengthen them as you solder the ends together.

3   File each corner of the rectangle slightly flat. Place it onto the soldered half-round wires and adjust the wires so they extend from each corner of the rectangle. Solder the two pieces together.

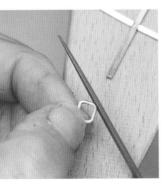

4   After pickling, rinsing, and drying, cut away the crossed area of the half-round wire, unless it is to be used as the attachment to the main piece.

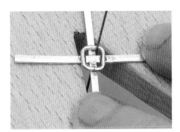

5   Now make another rectangle in exactly the same way, using 18-gauge (1-mm) round wire, but with outside measurements of ³/₈ in x ⁵/₁₆ in (10 mm x 6 mm). With a pair of pliers, bend up each corner of the frame from the first rectangle, until the second rectangle sits parallel to the first.

6   Solder in the second rectangle. Sit the stone in the frame to check the fitting. The claw wires should be true at each corner. At this stage, solder it to the main piece.

7   Make a mark where you want to cut each claw off, and pierce it away with the saw. File a small groove on the inside of the claw, where the girdle of the stone will be. Push all four claws down onto the stone.

Lesson 63: Making a cone

A cone-shaped bezel is sometimes made as the mount for a faceted stone. The cone follows the shape of the stone from girdle to culet. A cone is quite simple if you work with a formula. Take a look at the stone you want to set, and from that, decide the shape of the cone you need. You will need to know the diameter of the stone.

1   Draw the cone's side view, with the top measuring the diameter of the stone and the slope of the cone following the slope of the stone. Bring the slopes down to meet at a point which will be below the bottom of the stone. Mark off the height of the cone required on your drawing.

2   Take a compass, and open it out to the distance from the point to the top. Now draw an arc of about half a circle.

3   Close the compass up until it is level with the line marking the height of the cone, and draw another arc inside the first one.

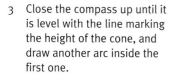

4   To find the length of the circumference needed, multiply the diameter of the top of the cone by 3.142 (π) and mark the length along the outer arc. Use a pair of dividers set at about one fifth of the length, and "walk" them along the arc five times.

5   Having marked the required length, take a pencil and ruler and draw a line from there, back to the original point.

6   Shade the area between the arcs and the lines. This shaded area is the shape of the metal that you need to cut to make the cone. Transfer it to the metal before piercing (see page 50).

Unit 33: Texturing

Texturing involves altering the smooth surface of metal to give it a distinct pattern or overall fabric-like quality. There are many different ways to achieve this look: some use a hammer, some use heat, some use little burrs, and others make good use of the rolling mill.

TEXTURING TIPS

Prepare the metal: Remove scratches by sanding with wet and dry papers, starting with 220 grit, then 320, 400, and finally 600 grit. Always texture metal that has been annealed. When using silver, anneal and pickle three times to make a soft, fine silver surface.

Use wooden tools: If the textured metal has to be shaped or bent in any way, use wooden, leather rawhide, or plastic mallets to avoid undermining the texture.

Take care with the rolling mill: If you are using the rolling mill to add texture, don't put anything through the mill that is harder than the steel rollers. If the hardness is questionable, use a protective sheet of 18-gauge (1-mm) copper between the material and the roller.

Work on a metal surface: When texturing with a hammer, hold the work on a metal surface to get the best results. The work may need to be annealed during the texturing process.

Hide solder lines: Solder lines can spoil a textured piece. If possible, place the solder where it is not going to run all over the pattern. (Soldering from the back is a good idea, too.) With a join on a textured ring, for example, there will be a visible solder line and there are likely to be signs of filing. However, if you place the ring on a mandrel and place some of the material you used to texture over the join, you can hit that with a metal hammer until you can't see the join anymore.

Texturing produces such an instant and wonderfully detailed effect that it has become extraordinarily popular. From the fragile outline of a feather to geometric patterns made using small pieces of another metal, the possibilities for imprinting designs with the help of the rolling mill are limitless. The surface of silver can be made to resemble the surface of the moon through a careful application of heat. Small burrs and cutters can be used with the flexshaft motor to produce lines and crisscrosses as well as a pattern of indentations.

Most texturing is done while the metal is still in sheet form and before cutting out has begun. This is because it usually involves some stretching of the metal, which would distort any piece that has been already cut to size.

TEXTURED STUDS
These stud earrings demonstrate how a fine all-over texture can add interest to an otherwise very simple design.

SILVER FINISHES
The silver surface of the brooch (right) was etched before the central decorative piece was applied. The silver surface of the pendant (left) was textured with a fine tool in the flexshaft motor.

Lesson 64: Texturing with the rolling mill

For this method use some thick watercolor paper with a rough texture.

1   Cut a 2 in x 1 in (60 mm x 30 mm) piece of 18-gauge (1-mm) silver sheet. Anneal and pickle the piece at least three times until it looks absolutely white, and dry it well. Offer it up to the rolling mill to see where it will start to grab it, and keep the mill in that position. There needs to be a good pressure on the metal for the texturing to be successful.

2   Tear an area out of the paper that is big enough to fold around both sides of the metal. Put the sandwich of the paper and the metal into the rolling mill and turn the handle to bring it all through to the other side. You may find this quite hard work!

3   Discard what is left of the paper—you will see that you cannot use these things twice—and have a look at the beautiful texture you have created.

TEXTURING BY ETCHING

Etching is the physical removal of metal by exposing it to a specific acid. For both copper and silver, you can use nitric acid, although ferric chloride is also suitable for use with copper. Nitric acid for etching is used in a solution of four parts water to one part acid. Always add the acid to the water, not the other way around. A glass or pyrex dish with a lid is used to contain the solution. Sometimes the nitric can be a little slow in "getting started." Place a little piece of scrap copper in the solution for a few minutes because this usually seems to start things moving. Use only plastic or stainless steel tweezers to get pieces in and out of the acid, and wear rubber gloves and a face mask of some sort. Try to have the dish in a ventilated area, such as under an open window.

The acid will only etch exposed areas of metal, so to achieve an interesting texture, asphaltum is painted onto the areas that are not to be etched. This of course must include the back of the metal and the sides. Alternatively, the piece of metal can be completely covered in asphaltum and, when dry, the pattern can be scratched through. When applying the asphaltum, you have to think slightly backward, to make sure you are letting the acid eat away in the right places!

ROLLING MILL TEXTURING TIP

Experiment with putting different materials through the rolling mill. Try using 240 grit wet and dry paper, for example, which gives a sparkly overall texture. Green leaves don't work well because they disappear into almost nothing, leaving a wet, soggy mess. Materials with different weaves, fine meshes, feathers, and so on, are all worth experimenting with to produce interesting textures. Anything harder than the steel rollers should be placed between copper sheets.

Lesson 65: Etching a strip of silver

Use this method to texture a strip of silver for use as a bangle.

1   Cut a 7 in x ⅝ in (180 mm x 20 mm) strip of 14-gauge (1.5 mm) silver. Anneal, pickle, and quench it. Test that the metal is completely clean by running it under the cold tap. If the water does not run off in little globules, the metal is clean; if it does, it is still greasy and needs annealing and pickling again.

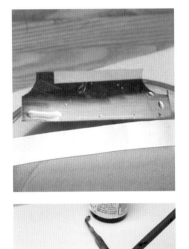

2   Handle the clean piece very carefully. If possible, hold it just by the edges and use a piece of kitchen paper to dry it. Lay the piece down so it is raised from the bench with a paper towel underneath it, and paint over all the back and sides with asphaltum.

3   When the asphaltum is dry, turn the piece over. Paint a line of asphaltum along the top, bottom, and side edges of the strip. This will prevent the acid eating away under the asphaltum painted on the edges. Now paint your pattern. Any areas you paint will remain raised and can be polished later, to contrast with the etched area.

4   Wait for the asphaltum to dry, and lower the strip into the nitric acid with a pair of tweezers. Put the lid on and check every 5 minutes. The metal will darken and you will see bubbling. Gently agitate the container to move the acid around.

5   When the etch is finished, the piece is ready to be shaped and finished as you choose.

PRACTICE PROJECTS

Use this technique to make the etched bangle; see page 130.

ETCHING TIPS

To check how deep the etch is, carefully lift the piece out of the acid and rinse it under cold running water. Pat it dry with a paper towel and check the edges between the asphaltum and the metal. You should see a small step. If not, put the piece back in the acid for a little longer. Once there is sufficient depth to the etch, remove it from the acid, rinse, and dry.

Place the piece on the soldering tray and heat it up until the asphaltum burns off. Clean it by scrubbing with a brass brush and liquid soap.

Lesson 66: Texturing with heat

There are two ways of texturing metal by heating it. The first is when enough heat is applied to start melting the surface (fusing) and the second is when the surface of the metal is moved by the central core of the metal starting to melt (reticulation). Both of these processes are completed before any soldering takes place. Because the amount of heat required is so close to the melting temperature of the metal, any solder already applied would burn out.

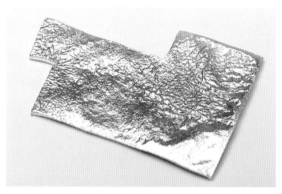

FUSED METAL
The piece of fused metal sheet, after thorough pickling.

RETICULATION

Use sheet silver with a thickness of more than 18-gauge (1 mm). Take a small sheet and bring the fine silver to the surface (see Lesson 21, page 59) through at least seven annealings and picklings. Once you have a fine white surface, the piece is ready. It is easier to achieve the movement you need if you work with two flames. Heat the piece of silver as if you were annealing it, and when the piece becomes faint red, introduce a second flame. This is a smaller, hotter flame. Concentrate it in one area, and as the surface starts to move, move the small flame to the next area. When you have worked across the piece, leave it to cool for a couple of minutes before quenching and pickling.

FUSING

Lay a piece of silver on the soldering or charcoal block. Introduce the flame and work the heat up the piece so it has a uniform heat throughout. Gradually increase the heat until you can see the metal starting to become a strong red color. Concentrate the heat in the red area until the surface of the metal starts to change in appearance and becomes almost shiny. You should see some light movement on the surface. Push the flame up until the next part of the metal starts to change, and continue to move the flame until the whole of the surface has been through the process. Remove the flame and wait a minute or two before quenching. The piece will be heavily oxidized so it will take longer than usual to clean in the pickle.

HEAT TREATMENT TIPS

Experiment first to get to know how the surface of the metal changes and just how far you can then take the heat before it all starts to melt. Here, silver is treated, but gold could also be used.

RETICULATED METAL
The clean piece of reticulated metal sheet, ready for filing and shaping.

Unit 34: Wirework

Many ideas and designs can be achieved by using wire alone, and it is an intrinsic part of many other aspects of jewelry making. Wire can be used with beads to make necklaces, earrings, and bracelets—and it doesn't always have to be soldered. It can be threaded through beads and made into links without having to go near any heat.

WIREWORK TIPS

Avoid marking wire: Wire is usually supplied in an annealed state. When bending it, use pliers that will not leave marks. Round/flat or half-round/flat are the best pliers to use, with the round side on the inside of the bend and the flat on the outside.

Neaten ends: After you cut lengths of thin wire, use a flat needlefile to neaten up the squeezed ends.

Making curves: When making curves, work with a piece of wire longer than you need. It is easier to get the correct curve by overcurving and then cutting away the excess than by trying to curve a short end.

Straightening wire: To straighten thin wire, anneal it, fasten one end in the serrated jaws of a vise, and hold the other end in a pair of serrated-edge pliers. Pull it out to give it tension and then give a quick but strong tug. To straighten a shorter length of thicker wire, hold the end in a pair of parallel pliers up to where there is a kink, and push it straight. Move that part of the wire down into the pliers and straighten the next area. Continue until the wire is straight.

Hardening wire: To harden a length of round wire, roll it along a steel flat plate or the top of an anvil, tapping it with a riveting hammer. To harden ear wires after they have been soldered, hold the pin as close as possible to the earring back with a pair of flat-nose pliers; then hold the end of the actual ear wire with another pair of flat-nose pliers and give them (and the wire) a 180° turn.

Wire comes in many different shapes and sizes. The most commonly used wire is round and supplied in sizes from 30-gauge (0.25 mm/cloisonné wire) to 2-gauge (6 mm), which would mostly be used for bangles and necklaces. It is a good idea to keep some wire in stock; the sizes you will find most useful are 20-gauge (0.8 mm), 18-gauge (1 mm), and 14-gauge (1.6 mm).

There is more about the different profiles of wire in Unit 5, page 21.

WIRE DETAIL
A delicate coil of gold wire around the single diamond setting complements the organic feel of this twiglike ring.

DECORATIVE BALLS
Beautifully made tiny gold balls adorn these gold and opal rings.

Lesson 67: Twisting wire

When different shaped wires are twisted together they become something altogether different from their original shapes. They can be made into substantial rings or form a delicate frame around a stone; they can be made into necklaces and intricate chains—it just takes a little imagination to achieve some highly individual pieces. To get you started, here is how to do a simple twist.

1   Coil up a 24-in (600-mm) length of 18-gauge (1-mm) round wire and place it on the soldering block. Anneal it carefully (see page 59). Quench, pickle, and dry.

2   Make a hook from an old wire coat hanger that you can fasten into a handled drill. Open out the coiled wire and fold it in half lengthwise. Place the two ends together in the serrated jaws of the vise.

3   Tighten up the vise to hold them securely. Pull gently down to the folded end and place it into the hook. Make sure the wire is not twisted at this stage.

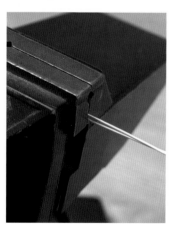

4   Wind the handle of the drill in a continuous smooth motion and watch the wire twisting itself around. Stop winding when you have a twist you like.

5   Remove the wire from both the hook and the vise and place it on the soldering block. Flux up the whole length and place small paillons of hard solder every ⅛ in (5 mm) or so. Run the solder. The wire is now ready for use and can be bent and worked, without opening.

WIRE-TWISTING TIP

Try twisting a square or triangular wire in with a round one, or add a plain round one in with the twisted one. When it is soldered up it can be hammered to give it a flat face or put through the rolling mill as gently or as hard as you like. Have a try and experiment!

Lesson 68: Wiring beads for a necklace

Wiring beads together to make a necklace is very satisfying, and knowing how to hang a bead off a necklace or earring will open up endless possibilities. Work with fine silver if possible because it will not oxidize and is very pliable, but if strength is required then use standard silver.

1   Most beads have a hole of $\frac{1}{32}$ in (0.5 mm) or slightly less. You will need to establish the hole size before starting, and if necessary pull down some 24-gauge (0.5-mm) wire in the draw plate (see page 104) so it fits through the hole. Start with a length of wire longer than you think you need. In one end of the wire, make a loop with a pair of round/flat pliers.

2   Bring the shorter length right past to make the loop round. Now bring the end back around the long length of wire and wrap it around it two or three times. Snip the remainder off as near to the wire as possible.

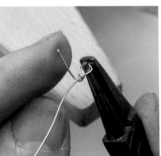

3   Thread a bead onto the wire. Where the wire exits the bead, push it down flat across the top of the bead.

4   Hold the wire with the round/flat pliers and form a loop as shown. Bring the end around as before. Snip the end of the wire off close to the loop and use a fine needlefile to smooth off the end.

5   Wrap the end around the wire that comes up through the bead, taking it down toward the top of the bead. Now make the link into the next bead. As you are making this bottom loop, thread it through the top loop of the first bead and then repeat the bead-threading process.

6   Make the loops big enough to allow a free movement of the chain, but not so big that they are out of proportion with the beads. Thread a jump ring and fastener onto the ends in the same way once you have strung a long enough length of beads for your piece.

Lesson 69: Making jump rings

Use this method to make jump rings with an outside diameter of approximately ¼ in (5 mm) with 18-gauge (1-mm) wire.

1   Anneal a 12-in (300-mm) long coil of wire. Quench, pickle, and dry it. Fasten a #31 (3-mm) steel former horizontally in the safe jaws of the vise. Make one turn of the wire around this former, and tighten it up together with the former in the vise with the rest of the wire being free. Hold the end of the wire and start wrapping it tightly around the former. Continue until all the wire is used.

2   To cut off the rings, work on the pin at the bench. Hold the coil on the former and slide the rings off the end one or two at a time, and cut through them one by one. The idea is to make sure the cut is neat and goes through the wire at a slight angle.

3   Let the rings fall into the skin or shelf beneath the bench. To prepare them for soldering, take a pair of flat-nose pliers in both hands and hold one side of the jump ring with the left-hand pliers. Take hold of the other side of the jump ring with the other pliers.

4   Bring the two ends together by twisting your hands together.

JUMP RING FORMING TIPS

Judging the size of the jump rings: When making jump rings, always make more than you need. Very tiny jump rings that are no bigger than around ⅛ in (3 or 4 mm) outside diameter can be made with 24-gauge (0.5-mm) wire, and anything going up to about ⁵⁄₁₆ in (7 or 8 mm) outside diameter could be made with 18-gauge (1-mm) wire. Jump rings that need to be bigger than this should increase in size proportionally. You will soon learn to judge whether a jump ring looks the right size in comparison with your work.

Soldering large numbers of jump rings: When making chains and other pieces that need many jump rings, solder as many as you can in one pass. Place them a little apart on your soldering block, with all the joins facing in the same direction. Flux each join, and then place a tiny chip of solder over it. Heat up each one individually, and move onto the next once you see the silvery shine of the solder flowing.

Lesson 70: Forming balls on the ends of wires

There will be many times when a rounded end to a length of wire will look much better than a blunt cutoff end. Designs that consist mostly of wire often have their ends neatened up by melting them into tiny balls. It may take a few attempts to get the technique right.

1   Hold a length of wire in a pair of insulated tweezers so about 1 in (25 mm) is exposed at the bottom. Dip the wire in flux, to keep it clean and allow it to pull into a bead easily. Hold the wire in front of a charcoal block. Bring a hot, fairly short flame onto the area just above the bottom of the wire, and play it there until you see the wire starting to curl up.

2   Let the ball travel up until it is about twice the diameter of the wire and then withdraw the flame. As the ball rolls up the wire, don't let it get any hotter because it is very easy to melt it all at this stage. Quench, pickle, and dry.

Lesson 71: Making balls

Small silver and gold balls can be very decorative on a piece of jewelry and are very simple to make.

1   Cut off ⅛-in (4-mm) lengths of 18-gauge (1-mm) round wire. Place the lengths of wire in indentations in the charcoal block, and use a hard hot flame on each one until it curls up into a ball.

2   Let each ball spin a for a second or two before moving on to the next one.

3   Once you have made all the balls, leave them to cool. To pickle them, place all the balls into a lidded plastic container, fill it with warm pickle, put the lid on, and let the balls sit for a while in the pickle bath until they are clean. In the meantime, dampen the charcoal block with a little water; otherwise it will keep burning.

Practice projects

As you work through this structured course, you will learn how to put new techniques into practice and create a selection of beautiful pieces. Each project has been specially designed to allow you to develop your skills progressively as you learn, and demonstrates how to marry great design with practical considerations such as choice of materials.

1

UNITS
12 • 14 • 15 • 16 • 18
21 • 22 • 24 • 32

Cabochon-set ring

(page 126)
This project features a silver ring with a cabochon stone in a rub-over setting. You will need to find the correct length for both the ring band and the bezel for the stone. The band can be any width you like, but the bezel setting for the stone should not be any wider than the band.

TECHNIQUES
Unit 12 Measuring
Unit 14 Piercing
Unit 15 Annealing
Unit 16 Quenching and pickling
Unit 18 Bending
Unit 21 Soldering
Unit 22 Filing
Unit 24 Polishing and finishing
Unit 32 Stone setting

2

UNITS
14 • 16 • 21
22 • 27 • 28

Cast cuff links

(page 129)
These attractive cuff links are made using the cuttlefish casting method to produce interesting surfaces. You can design your own patterns by carving them directly into the bone, or press found objects, such as shells, into it.

TECHNIQUES
Unit 14 Piercing
Unit 16 Quenching and pickling
Unit 21 Soldering
Unit 22 Filing
Unit 27 Fittings
Unit 28 Casting

3

UNITS
12 • 14 • 15 • 16 • 17
18 • 21 • 22 • 33

Etched bangle

(page 130)
Making an open bangle is a simple and very effective way of making a bracelet which does not require an exact measurement. The proportions of the bangle shown here can be varied. The etched decoration can be adapted to your own design, or you could use a different method of texturing.

TECHNIQUES
Unit 12 Measuring
Unit 14 Piercing
Unit 15 Annealing
Unit 16 Quenching and pickling
Unit 17 Cleaning
Unit 18 Bending
Unit 21 Soldering
Unit 22 Filing
Unit 33 Texturing

4
UNITS
14 • 15 • 19
21 • 22 • 24

Wire-decorated earrings

(page 132)
Making this pair of earrings provides
excellent soldering practice as it uses
hard, medium, and easy solders. For these
earrings, 24-gauge (0.5-mm) gold wire
was used as decoration, but you could
use silver wire instead.

TECHNIQUES
Unit 14 Piercing
Unit 15 Annealing
Unit 19 Cutting
Unit 21 Soldering
Unit 22 Filing
Unit 24 Polishing and finishing

5
UNITS
14 • 15 • 16 • 17
21 • 22 • 31 • 33

Textured pendant

(page 134)
This pendant was textured by putting it
through the rolling mill with a piece of
fabric; in this case lace, which produces
a strong pattern.

TECHNIQUES
Unit 14 Piercing
Unit 15 Annealing
Unit 16 Quenching and pickling
Unit 17 Cleaning
Unit 21 Soldering
Unit 22 Filing
Unit 31 Using a rolling mill
Unit 33 Texturing

6
UNITS
12 • 14 • 15 • 16 • 18
21 • 22 • 24 • 26 • 27

Circular brooch

(page 136)
The design of this brooch shows off two
complementary finishes, and cutting the
circles will test your piercing skills. This
project also includes a yellow and white gold
decoration, which must be soldered on in a
particular way.

TECHNIQUES
Unit 12 Measuring
Unit 14 Piercing
Unit 15 Annealing
Unit 16 Quenching and pickling
Unit 18 Bending
Unit 21 Soldering
Unit 22 Filing
Unit 24 Polishing and finishing
Unit 26 Drilling
Unit 27 Fittings

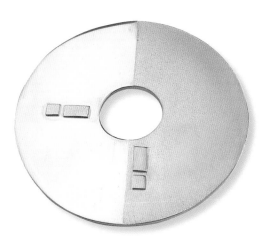

SEE ALSO
Unit 12, Measuring, page 46
Unit 18, Bending, page 64
Unit 32, Stone setting, page 108

Project 1: Cabochon-set ring

This project features a silver ring with a cabochon stone in a rub-over setting. You will need to find the correct length for both the ring band and the bezel for the stone. The band can be any width you like, but the bezel setting for the stone should not be any wider than the band.

MATERIALS:
14-gauge (1.5-mm) silver sheet
18-gauge (1-mm) silver sheet
24-gauge (0.5-mm) gold or
  silver sheet
Flux
Silver solder: hard and easy
Pickle
Flat-backed cabochon stone

TOOLS:
Dividers
Ruler
Saw
Half-round pliers
Soldering block or firebrick
Torch
Paper towels
Files
Round mandrel
Mallet
Parallel pliers
Binding wire
Wet and dry papers
Blue tack
Handheld clamp
Pusher, or punch and hammer
Burnisher

1   Open the dividers to the width of the band required for the ring. Run them down the edge of the 18-gauge (1-mm) silver sheet, to the length required. (See page 47 for calculating the length.) Pierce along the line. Anneal the strip, quench, and pickle.

2   Use a pair of half-round pliers, with the half-round side on the inside curve of the silver, to bend up the ring ready to be soldered.

3   Place the ring on the soldering block and flux the join. Place a little paillon of hard solder under the join and solder it up. Pickle, rinse, and dry with paper towels. File away any excess solder and round up the band on the mandrel.

4   Measure the length and width of the stone with a pair of dividers. Use the measurements to calculate the length and width of the bezel (see page 111).

5   Mark the width and length of the bezel with a pair of dividers onto a 24-gauge (0.5-mm) sheet of gold or silver.

6   Pierce out the strip of silver or gold and anneal it. Pickle, rinse, and dry. Solder the join with hard solder. (Use gold solder with gold and silver with silver.) File away excess solder, and use a mallet and a mandrel to round up the bezel. Squeeze it in a pair of parallel pliers until it is the correct shape for the stone.

7 Place the bezel onto some 14-gauge (1.5-mm) silver sheet. Flux around the join and use hard silver solder to solder it on. When soldering gold to silver, use silver solder.

8 After soldering, use the saw to pierce away excess metal from around the bezel. File around the outside edge, so the solder line disappears.

9 The bottom of the setting must be made to fit the ring. Place it on an oval or half-round file with a similar profile to the ring band. Keep it straight and file it down.

10 File the setting until it fits the top of the ring.

11 Hold the setting and the ring band together with binding wire, ready for soldering.

12 Place the ring onto the soldering block. Flux and place paillons of easy solder around the join, then bring the flame gently onto the base of the ring. Gradually increase the heat all around until the solder melts. After soldering, pickle, rinse, and dry. Do any filing necessary and clean the ring with graded wet and dry papers before setting the stone.

13 Put a little blue tack on top of the stone. (This helps to lift it in and out of the setting.) File down the top of the bezel until the stone sits correctly in it. (The bezel should come to the part of the stone that just starts to curve over. If the bezel is too high it is more difficult to set the stone.)

14 Hold a flat needlefile at 45° to the top edge of the bezel, and file all around it to thin it a little.

15 Place the ring into a handheld clamp. Place the stone into the setting, and use the pusher to start pushing over the top edge. Push one side, then turn the clamp around and push the opposite side. Continue pushes from opposite sides, until the bezel sits neatly around the stone.

16 Alternatively, if the bezel feels too hard to push over, place the clamp and ring into a vise and use a small punch and hammer to push it over. Again, work from opposite sides. Hold the polished end of the punch on the top edge of the bezel and tap quite gently.

17 When the bezel is sitting neatly around the stone, you should not be able to see any gaps. If necessary, clean it up with the edge of a flat needlefile, with your thumb held over the stone to protect it. Use the burnisher to polish the edges.

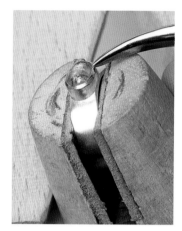

FINISHED RINGS
These rings demonstrate how you can take the basic technique and vary the profile of the shank, the height and thickness of the setting, and the type and cut of the stone, to achieve an almost endless variety of effects.

Project 2: Cast cuff links

These attractive cuff links are made using the cuttlefish casting method to produce interesting surfaces. You can design your own patterns by carving them directly into the bone, or press found objects such as shells into it. This project uses the silver strip made in Lesson 50, page 97. The cuff link fittings shown here can be soldered on complete.

SEE ALSO
Unit 27, Fittings, page 92
Unit 28, Casting, page 96

MATERIALS:
Silver casting grains
Cuttlefish bone
Pickle
Cuff link fittings
18-gauge (1-mm) silver wire
Easy solder

TOOLS:
Saw frame and thick blade
Files
Torch
Soft brass brush and liquid
   soap or polishing machine

1   Cast the silver strip as shown in Lesson 50, page 97. Pickle it and pierce away the sprue. Pierce the shapes for the cuff links from the length of silver. Follow the curved lines of the pattern on one end of the piece, and cut straight across for the other end.

2   File away the sharp corners and edges; then smooth the bottom with a large flat file.

3   To make the cuff links a little lighter, use a flat file to remove some of the metal from underneath in a curve up to the top edge.

4   The fittings used here have holes in the bottoms. Use these to locate them on the center back of the cuff link. Solder a piece of 18-gauge (1-mm) silver wire onto the back with hard solder.

5   Place the fitting over the pin and use easy solder to solder it on. Cut away the excess pin length from the center of the fitting and file it level.

**THE FINISHED
CUFF LINKS**
The cuff links were finished by brushing with a soft brass brush and liquid soap.

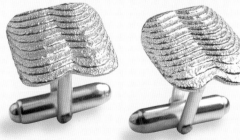

SEE ALSO
Unit 12, Measuring, page 46
Unit 33, Texturing, page 115

Project 3: Etched bangle

Making an open bangle is a simple and very effective way of making a bracelet that does not require an exact measurement. As it is put on and taken off the wrist, it will be subject to a certain amount of stress, and so the sheet metal you use does need a thickness of at least 18-gauge (1 mm). Solder the bracelet together and then shape it around the mandrel, before opening it up, to achieve a good shape.

MATERIALS:
Silver sheet, at least
　18-gauge (1 mm) thick
Asphaltum
Nitric acid solution
　(4 parts water:1 part acid)
Pickle
Flux
Silver solder

TOOLS:
Saw
Two pieces of wooden dowel
Small paintbrush
Plastic or stainless steel
　tweezers
Plastic container
Paper towels
Torch
Files
Firebrick
Charcoal blocks
Oval mandrel
Mallet
Wet and dry papers
Soap-filled fine steel-wool pad

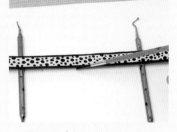

1　Saw a ¾ in x 7 in (20 mm x 180 mm) strip of 18-gauge (1-mm) silver. Leave a little extra around all the edges. Anneal, rinse, and dry the strip, without touching it with your fingers. Lay it across two hand tools or wooden dowels, and then paint the edges of the strip and your chosen pattern onto the surface with the asphaltum.

2　When the top is dry (around 20–30 minutes), turn the strip over and cover the back with asphaltum. Allow it to dry.

3　Use a pair of plastic or stainless steel tweezers to pick up the strip, and place it in a plastic container of nitric acid solution. Tilt the plastic container if necessary until the strip is completely covered by the acid solution. Every 5 minutes, tip the plastic container to move the liquid over the strip. The etching should take around 20 minutes.

4　Pick up the strip with the tweezers and rinse it under a cold tap. Gently pat it dry with kitchen paper. If the etching is not yet deep enough, put the strip back into the acid for a few more minutes. When it is deep enough, anneal the strip to burn off the asphaltum. Pickle, rinse, and dry.

5　Bend the strip around until the two ends meet. Push them past each other and then bring them back to sit together. If they do not meet neatly, place a flat file between the two ends and file them until they do fit together.

6   Place the bangle on a soldering block or firebrick. Make an "oven" by placing charcoal blocks around the sides and back of it to help concentrate the heat.

7   Flux the join and place three or four small paillons of hard solder across the back of the join. Use a large flame to solder the join. Pickle, rinse, and dry. Carefully file any excess solder from the front and underside of the join.

8   Shape the bangle around an oval mandrel with a leather rawhide or wooden mallet. You can hit the bangle quite hard with the mallet, which will put some tension into the silver.

9   Cut through the join with a saw. It should spring apart.

10 Open up the bangle by placing it back onto the mandrel and tapping it with the mallet until it is the correct size.

11 Use a file to round the edges of the opening so the bangle is comfortable to put on and take off. Do any necessary filing and cleaning around the edges.

12 Clean the inside of the bangle with some wet and dry paper. The bangle can now be finished using a series of wet and dry papers, or simply run under the cold water tap and rub gently with a soap-filled fine steel-wool pad.

THE FINISHED BANGLE
This example was finished with fine steel wool and liquid soap. The edges were smoothed and slightly rounded with a needlefile.

Project 4: Wire-decorated earrings

Making this pair of earrings provides excellent soldering practice because it uses hard, medium, and easy solders. For these earrings, 24-gauge (0.5-mm) gold wire was used as decoration, but you could use silver wire instead.

SEE ALSO
Unit 21, Soldering, page 72
Unit 34, Wirework, page 119

MATERIALS:
8-gauge (3 mm x 1.5 mm)
    D-section wire
24-gauge (0.5-mm) silver or
    gold round wire
Flux
Solder: hard, medium, and easy
Butterfly and pin earring fittings

TOOLS:
Saw
Torch
Side or top cutters
Files
Soldering block or firebrick
Small paintbrush
Wet and dry papers
Tweezers
Fine steel-wool pad and
    liquid soap (optional)

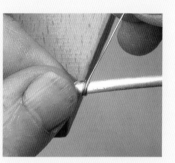

1   Anneal the 24-gauge (0.5-mm) wire. (It may not need annealing. Try bending it gently to check.) Hold one end of the wire against the D-section with your fingers. With the other hand, wrap the thinner wire as tightly as possible around the D-section wire. Wrap it around at least eight times and slide it off.

2   Cut two 1-in (25-mm) pieces from the D-section wire. Use the saw to cut the ends at an angle. Cut two ⅝-in (15-mm) pieces, with a similar angle at top and bottom.

3   Use a pair of side or top cutters to snip open each piece of thin wire, along the flat side of the profile.

4   Push two pieces of the thin wire onto the longer pieces of the D-section. Push one piece onto each of the shorter pieces. The wires should line up when you solder the D-sections together. Flux around the wire, place paillons of hard solder at each edge, and solder. Try not to solder the wire to the flat bottom.

5   Next, prepare to solder the two pieces of D-section wire together. First, file away the excess wire from the sides that will fit together.

6   Use the side or top cutters to clip away the wire from the flat side of the D-section.

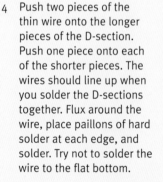

7   File it so that none remains.

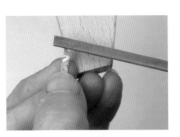

8   Place the two different lengths of D-sections together on the soldering block so the decorative wires line up, and run a little flux down the middle. Place tiny paillons of medium solder at each end and one or two in the middle. Solder together. Clean away excess solder with a fine file and use some 400-grit wet and dry paper to clean the back.

9   Place the earring upside down on the soldering block or firebrick. Put a little flux where the pin is to go, and place a paillon of easy solder onto it. Hold the pin in a pair of tweezers with one hand. Heat up the earring using the torch in your other hand, and when the solder starts to run, lower the pin into the puddle of solder. Hold it for a second or two and then remove the flame. Hold the pin for a few seconds more; then lift the whole earring up and place it into the pickle.

10   Repeat steps 3 through 9 for the second earring and then, to achieve a matte finish, use wet and dry papers, and possibly some fine steel wool with liquid soap. Fit the butterfly fittings onto the earring posts.

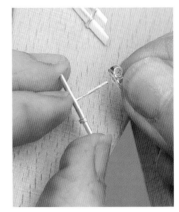

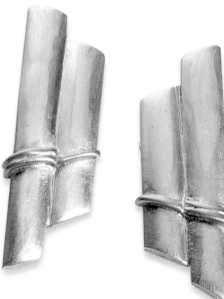

THE FINISHED EARRINGS
Combining silver with gold decoration, as in these elegant earrings, adds real interest to a piece.

Project 5: Textured pendant

This pendant was textured by putting it through the rolling mill with a piece of fabric. You can adapt the technique to use most types of fabric or even watercolor paper. In this project, lace was used, which makes a very strong pattern to great effect.

SEE ALSO
Unit 15, Annealing, page 58
Unit 31, Using a rolling mill, page 106
Unit 33, Texturing, page 115

MATERIALS:
Piece of 16-gauge (1.2-mm) silver sheet approximately 3^1/$_8$ in x 2 in (80 mm x 50 mm)
3 pieces of silver tubing approximately ^{1}/$_8$ in (3 mm) in diameter and ^{1}/$_8$ in (3 mm) long
Textured fabric or paper
Pickle
Flux
Solder: hard and easy
Sheet of mica

TOOLS:
Torch
Soldering block or firebrick
Paper towels
Rolling mill
2 flat pieces of metal
Hammer
Ruler
Mechanical pencil
Files
Wet and dry papers
Fine steel-wool pad and liquid soap
Impregnated silver cloth

1   Anneal the silver at least three times (see page 59). Each time, leave the silver in the pickle for at least 5 minutes. At the end of this process the resulting layer of fine silver will take an impression well.

2   Cut out a piece of fabric or paper larger than the silver. Open the rollers until they grip the silver. Remove it. Turn the rollers down a fraction and place the fabric on top of the silver. Hold both up to the rollers and roll them through.

3   The silver should now have a good impression in it. It will be a little thinner and may need straightening out. Anneal it again; then place it between two flat metal sheets with a piece of the same fabric on top of it.

4   Use a hammer to tap the top sheet down gently until the silver inside is flattened.

5   The sheet is now ready to be cut into three pieces. Use a ruler to help you draw two straight lines at an angle down the patterned side of the silver.

6   Now cut out the three pieces of the pendant with the saw. File the edges of each piece and lay them against each other to check the fit. Here, the two outer pieces were cut shorter than the middle one.

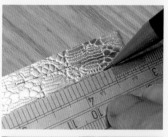

7 Use hard solder to solder the small pieces of tubing onto the back of each piece. Make sure they line up across the three pieces in the position that you want them on the chain.

8 Remove any excess solder with a fine file, and clean the backs with wet and dry paper. Clean the fronts with steel wool and liquid soap.

9 To fit a standard snake chain, carefully cut through the loop with the saw where it meets the chain. Use as fine a blade as possible.

10 Use a pair of round/flat pliers to straighten out the loop.

11 Thread the chain through each piece of tubing.

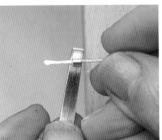

12 Bend the loop so it fits snugly back into position. Place it onto the soldering block with the pendant pieces at the other end of the chain. Flux the join and place a paillon of easy solder across it.

13 Place a sheet of mica across the chain, so the end being soldered is the only area to have the flame directly on it. When the soldering is finished, hold the end of the chain in the pickle to clean it; then rinse and dry. Clean the end with an impregnated silver cloth.

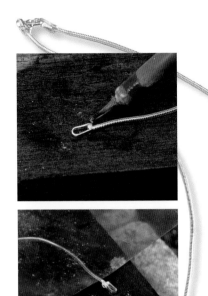

THE FINISHED PENDANT
The pattern in the lace allowed a flower to form the central focus of the three-piece pendant.

Project 6: Circular brooch

The design of this brooch shows off two complementary finishes, and cutting the circles will test your piercing skills. This project also includes a yellow and white gold decoration, which must be soldered on in a particular way.

MATERIALS:

18-gauge (1-mm) silver sheet, 3½ in (90 mm) square
Silver tubing ¹⁄₁₆ in (1 mm) inside diameter, approximately ½ in (10 mm) long
18-gauge (1-mm) wire 4 in (100 mm) long
12 x 18-gauge (2 mm x 1 mm) D-section wire approximately 3¼ in (80 mm) long
Lengths of 24-gauge (0.5-mm) white and yellow (18-carat) gold wire
Flux
Solder: hard and easy

TOOLS:

Dividers
Saw
Punch
Hammer
Drill
Files
Scribe
Masking tape
Polishing machine and stainless steel wheel
Small paintbrush
Torch
Soldering block
Paper towels
Flat needlefile
Wet and dry papers
Round/flat pliers
Parallel pliers
Top or side cutters

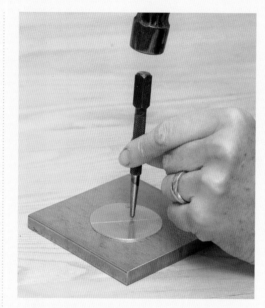

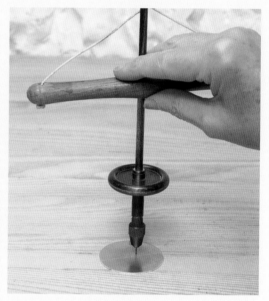

1   Use a pair of dividers or a template to draw a circle with a diameter of 3¼ in (80 mm) on the silver sheet. Mark the outside edge and the edge for the inner circle. Make sure that the point of the dividers is firmly on the silver to then draw the circles with the other point. You may find cutting the circle easier if you then open out the dividers a fraction and draw a second circle. Use the space between the two lines to cut out the circle. Use a center punch to mark where you will drill the hole for the inner circle.

2   Drill the hole. (Making a hole at the edge of the inner circle is less wasteful than making a hole in the center.)

SEE ALSO
Unit 14, Piercing, page 52
Unit 24, Polishing and finishing, page 84
Unit 27, Fittings, page 92

3. Thread your saw blade through the hole, and cut out the inner circle.

4. File the inner circle to the scribed line with an oval or half-round file.

5. File the outside edge down to the scribed line with a flat file.

6. Find the halfway line across the circle, and use the point of a scribe to mark it.

7. Stick a piece of masking tape firmly along the scribed line.

8. Fit the stainless steel wheel onto the polishing machine, and use it to texture the uncovered side of the silver.

9. Cut some tiny rectangles of yellow and white gold with the saw. Make sure they are completely flat. On the back of each, place some flux and a chip of hard solder, and heat them up until the solder runs. Pickle, rinse, and dry the pieces; then use a flat needlefile to file the "bump" of solder nearly flat.

10 Stick masking tape over the textured side of the silver, and polish the exposed side with wet and dry papers, from 400 grit down to 1200 grit, to get a smooth and scratch-free finish.

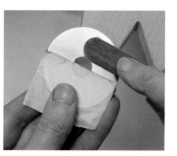

11 Paint flux onto the back of the gold pieces, and place them on the brooch. Heat it up gently so the gold pieces stay in place and concentrate the heat on the silver while watching for the silver line as they solder onto the piece.

12 To make the pin, anneal the 18-gauge (1-mm) round wire. Use a pair of round/flat pliers to form a tight circle at one end. Continue turning the wire around the pliers to make a spiral.

13 It helps to keep the spiral flat if you hold it as you turn with a pair of parallel pliers. Make about two whole turns around the center.

14 Use the D-section wire to make the catch. With the round side of the round/flat pliers on the curved top of the D-section, make a loop for the hook.

15 Push the flat side so it sits straight. Cut the end off the flat section of wire with top or side cutters, leaving about 3/16 in (5 mm). The hook should balance on the flat section.

16 Place the hook and the tubing on the back of the brooch. Flux around the edges, and place two chips of hard or medium solder against each. Heat the piece very gently, keeping the flame near the middle area so it doesn't touch either the hook or the tubing. Keep the flame moving until the solder runs. Pickle, rinse, and dry.

17 Thread the pin through the tubing so the spiral end just sits underneath it. Flux all around that end, and place paillons of easy solder around and inside the spiral. Heat up the brooch gently, keeping the flame away from the wire pin, until the solder runs. Pickle, quench, and dry.

19 Cut off the end and file to a point. Smooth it by rubbing hard with the burnisher.

18 Bend the pin as close to the tubing as possible with a pair of small, flat-nose pliers so it fits into the catch.

THE FINISHED BROOCH

For this brooch, the the decorative pieces of white gold soldered to the polished half of the brooch were textured. The pieces of yellow gold soldered to the textured half were polished.

Conversion charts

METAL THICKNESSES
Precise metric and imperial gauge conversions

B&S gauge	mm	inches
2	6.54	0.258
4	5.19	0.204
6	4.11	0.162
8	3.26	0.129
10	2.59	0.102
12	2.05	0.081
14	1.63	0.064
16	1.29	0.051
18	1.02	0.040
20	0.81	0.032
22	0.64	0.025
24	0.51	0.020
26	0.40	0.015
28	0.33	0.013
30	0.25	0.010
32	0.20	0.008
34	0.16	0.006
36	0.13	0.005

QUICK REFERENCE MM/GAUGE SIZES

mm	nearest B&S gauge
3.00	8
2.50	10
2.00	12
1.50	14
1.25	16
1.00	18
0.75	20
0.64	22
0.50	24

RING SIZES

US	UK	Europe	mm	inches	US	UK	Europe	mm	inches	US	UK	Europe	mm	inches
½	A		37.83	1.490	5	J ½	9	49.20	1.938	9 ½	T	21	61.13	2.408
¾	A ½		38.42	1.514	5 ¼	K	10	49.80	1.962	9 ¾	T ½	22	61.77	2.434
1	B		39.02	1.537	5 ½	K ½		50.39	1.986	10	U		62.40	2.459
1 ¼	B ½		39.62	1.561	5 ¾	L	11	50.99	2.009	10 ¼	U ½	23	63.04	2.484
1 ½	C		40.22	1.585	6	L ½		51.59	2.033	10 ½	V	24	63.68	2.509
1 ¾	C ½		40.82	1.608	6 ¼	M	12	52.19	2.056	10 ¾	V ½		64.32	2.534
2	D	1	41.42	1.632	6 ½	M ½	13	52.79	2.080	11	W	25	64.88	2.556
2 ¼	D ½	2	42.02	1.655	6 ¾	N		53.47	2.107	11 ¼	W ½		65.48	2.580
2 ½	E		42.61	1.679		N ½	14	54.10	2.132	11 ½	X	26	66.07	2.603
2 ¾	E ½	3	43.21	1.703	7	O	15	54.74	2.157	11 ¾	X ½		66.67	2.627
3	F	4	43.81	1.726	7 ¼	O ½		55.38	2.182	12	Y		67.27	2.650
	F ½		44.41	1.750	7 ½	P	16	56.02	2.207	12 ¼	Y ½		67.87	2.674
3 ¼	G	5	45.01	1.773	7 ¾	P ½		56.66	2.232	12 ½	Z		68.47	2.680
3 ½	G ½		45.61	1.797	8	Q	17	57.30	2.257					
3 ¾	H	6	46.20	1.820	8 ¼	Q ½	18	57.94	2.283					
4	H ½		46.80	1.844	8 ½	R		58.57	2.308					
4 ¼	I	7	47.40	1.868	8 ¾	R ½	19	59.21	2.333					
4 ½	I ½	8	48.00	1.891	9	S	20	59.85	2.358					
4 ¾	J		48.60	1.915	9 ¼	S ½		60.49	2.383					

Index

Resources

MAGAZINES

Lapidary Journal Jewelry Artist
300 Chesterfield Parkway, Suite 100
Malvern, Pennsylvania 19355
Tel: 610-232-5700
Fax: 610-232-5756
www.jewelryartistmagazine.com

Crafts
44a Pentonville Road
London N1 9BY
United Kingdom
Tel: +44 (0) 20 7806 2542
Fax: +44 (0) 20 7837 0858
www.craftscouncil.org.uk

Metalsmith
Society of North American Goldsmiths
540 Oak Street, Suite A
Eugene, Oregon 97401
Tel: 541-345-5689
Fax: 541-345-1123
www.snagmetalsmith.org

Retail Jeweller
33–39 Bowling Green Lane
London EC1R 0DA
United Kingdom
Tel: +44 (0) 20 7812 3724
Fax: +44 (0) 20 7812 3720
www.retail-jeweller.com

BOOKS

Crowe, Judith
Jeweller's Directory of Gemstones
A&C Black, 2006

Fisher, Mark
Britain's Best Museums and Galleries
Allen Lane, 2004

Haab, Sherri
The Art of Metal Clay (with DVD)
Watson-Guptill, 2007

McCreight, Tim
Jewelry: Fundamentals of Metalsmithing
Hand Books Press, 1997
Complete Metalsmith
Brynmorgen Press, 2004

Olver, Elizabeth
Jewelry Design: The Artisan's Reference
North Light Books, 2000
The Art of Jewelry Design:
From Idea to Reality
North Light Books, 2002

Untracht, Oppi
Jewelry Concepts and Technology
Doubleday, 1982

Van de star, Renee
Ethnic Jewellery
Pepin Press, 2006

WEBSITES

www.jewelrymaking.about.com
www.snagmetalsmiths.org
www.whoswhoingoldandsilver.com
www.jaa.co.au
www.acj.org.uk

Credits

For the new ones.

Thank you so much to all my fellow jewelers who supplied photos of their work for this book. Their names are supplied below. Thank you to HS Walsh of Beckenham and London for the loan of all the new tools and to Prabhu Enterprises and Affinity Gems for their generous loan of stones and beads. A big thank you to all my students who, over the years, have taught me so much. Thanks too to Peta from Kings Framers for her pictures in the "museum." And lastly thank you Paul, for fun photography, and Liz for making it all stress free!

Jinks McGrath

Quarto would like to thank the following artists, (and photographers in brackets), for kindly submitting images for inclusion in this book:

Key: a = above, b = below, c = center, l = left, r = right

Elaine Cox www.elainecox.co.uk 54b, 64c, 101a; Robert Feather www.robertfeather.co.uk 64a; John Field www.jfield.co.uk 49br, 108br; Shelby Fitzpatrick www.shelbyfitzpatrick.com (Mike Blissett) 64r, 83r, 92b; Clare Ford 49r, 72ar; Charmian Harris www.charmianharris.co.uk 64bl; Rauni Higson www.raunihigson.co.uk 17ar, 17br, 48ar, 67a, 67b, 80ar, 80br; Jon and Valerie Hill 5—2nd from right, 9a, 18r, 46, 64cl, 108al, 112ar, 115a, 143; Karen Holbrook 140, 141; Ulla Hörnfeldt www.ullahornfeldt.com 40al, 72br; Daphne Krinos www.daphnekrinos.com (Joël Degen) 85ar; Linda Lewin www.lindalewin.co.uk 119r; Jane Macintosh www.janemacintosh.com (Joël Degen) 8al, 8br, 34r, 70al, 76r; Jane Macintosh www.janemacintosh.com (Full Focus) 85br; Catherine Mannheim www.catherinemannheim.com (FXP) 34l; Al Marshall www.fluxnflame.co.uk 56l; Jesa Marshall www.fluxnflame.co.uk 14l, 90a; Jinks McGrath 90ar; Guen Palmer www.guenpalmer.com (Full Focus) 92a, 108ar, 112br; Nicola Palterman www.nicolapalterman.com 115b, 119l; Kate Smith www.katesmithjewellery.co.uk 16; Susy Telling www.susytelling.com 42r; Mari Thomas www.marithomas.com 21bl, 44; Alan Vallis www.alanvallis-oxo.com 96, 108bl; H.S.Walsh & Sons www.hswalsh.com 106.

All other images are the copyright of Quarto Publishing plc. While every effort has been made to credit contributors, Quarto would like to apologize should there have been any omissions or errors—and would be pleased to make the appropriate correction for future editions of the book.